I0114535

.

STRESS, SUICIDES AND FRATRICIDES IN THE ARMY-
CRISES WITHIN

This book is in honour of all those gallant officers and men who have been or are engaged in an extremely complex conflict; yet survived the adverse impact of all the obscure situations.

STRESS, SUICIDES AND FRATRICIDES IN THE ARMY-
CRISES WITHIN

Maj Gen Samay Ram, UYSM, AVSM,VSM (Retd)

(Established 1870)

United Service Institution of India
New Delhi

Vij Books India Pvt Ltd
New Delhi (India)

Published by

Vij Books India Pvt Ltd
(Publishers, Distributors & Importers)
2/19, Ansari Road, Darya Ganj
New Delhi - 110002
Phones: 91-11-43596460, 91-11- 65449971
Fax: 91-11-47340674
e-mail : vijbooks@rediffmail.com

Copyright © 2011, United Service Institution of India, New Delhi

First Printed   : 2011
Second Reprint   : 2012

ISBN: 978-93-80177-41-0

All rights reserved.

No part of this book may be reproduced, stored in a retrieval system, transmitted or utilized in any form or by any means, electronic, mechanical, photocopying, recording or otherwise, without the prior permission of the copyright owner. Application for such permission should be addressed to the publisher.

Contents

LIST OF APPENDICES

LIST OF TABLES

Foreword

It gives me great pleasure to know that the book titled 'Stress, Suicides and Fratricides in the Army –the Crises Within', authored by Maj Gen Samay Ram, UYSM, AVSM, VSM (Retd) is going to be published soon.

The Indian Army has a glorious history and tradition of service to the Nation. The gallant officers and men, engaged in extremely complex conflicts since our Independence, have bravely faced all the challenges. The supreme sacrifice of our valiant soldiers has left indelible marks in the hearts of our Nation and its citizens. Our country is secure in their hands and is today, on the path to becoming one of the major powers of the world. However, at the same time, when we lose brave officers and men not at the hands of the adversary or in battle, but due to stress, fatigue or tensions, it becomes a cause of concern, not only for the Army authorities and the Ministry of Defence, but also for the entire country. Have we failed to live up to and perform our duty towards those who are serving the Country with utmost sincerity and dedication?

This study, is no doubt, a laudable effort with an aim of unfolding the causes and evolving a solution to the syndrome of suicides and fratricides taking place amongst all ranks of the Armed Forces. The author has undertaken a deep and analytical study on management of stress in the Army, and has carried out an extensive research to highlight the root causes that influence the psyche of the officers and men when placed under trying conditions. He has presented a fairly comprehensive picture of the scenario, through a scientific analysis combined with practical inferences.

The book provides a good insight into not only the Army but the society as a whole, and highlights the changes that have taken place in the socioeconomic milieu, resulting in changed societal patterns and value systems which have also contributed to the problem of stress. Core dimensions of the service conditions and problems confronted by the solider, and their evolution over a period of time have been dealt with comprehensively.

I commend the book with the hope that it would be a valuable referral document not only for the higher leadership and officers of the Defence Services, various Para-Military and other Security Forces, but also for policy makers, administrators, social scientists, academicians and readers in general. I trust that it will be beneficial to all and I convey my sincere appreciation of the efforts put in by Maj Gen Samay Ram, in bringing out this valuable composition.

Raj Bhavan   General JJ Singh
Itanagar   PVSM, AVSM, VSM (Retd)
Arunachal Pradesh   22 August 2010

Acknowledgement

I wish to acknowledge my debt of gratitude to all those people and institutions that helped me in taking up and completing this study work. I am particularly indebted to Lt Gen Satish Nambiar, PVSM, AVSM, VrC (Retd); the erstwhile Director of USI for being very constructive in his suggestions. I also thank Lt Gen P K Singh, PVSM, VSM (Retd), Director USI for giving his most valuable time and views which helped me in projecting a balanced approach to the problem.

I am deeply indebted to HE Gen J J Singh, PVSM, AVSM, VSM (Retd); Governor of Arunachal Pradesh who gave his blessings. There are also a host of retired officers who not only promptly answered my questionnaire but also gave their valuable time for discussion. I have benefited immensely from the experience of some of the finest officers; prominent among them being Lt Gen R K Nanavatty, PVSM,UYSM,VSM(Retd) who is known for his incisive comments- Lt Gen V G Patankar, PVSM, AVSM,VSM (Retd); Lt Gen V K Singh, PVSM (Retd), Lt Gen Y M Bammi, PVSM, AVSM, VSM(Retd); and Maj Gen Dalvir Singh, AVSM, VrC, VSM (Retd); for his extensive hand written notes and for giving his perception at the ground level.

The list is long but I cannot conclude without mentioning my deep gratitude to Lt Gen J B S Yadava, PVSM AVSM, VrC, VSM (Retd); who interacted regularly and contributed his views, based on his most valuable experience as a Formation Commander, at various levels.

There are many whose names cannot be mentioned but whose views could not be ignored because they are the ones who are directly faced with the challenge of managing the complex problems. I salute these brave officers and men who despite very adverse circumstances are successfully accomplishing their tasks

Lastly, my thanks to all those who helped me in preparing the Slides/Graphs and giving shape to the book.I would fail in my duty if I do not acknowledge the support given to me by my wife. I thank her for bearing with the nuisance of my disturbing her peace at home.

- Author

Introduction

In my perception based on my experience of over three decades service including almost seven years abroad; four in Afghanistan as the Military Attaché of India in Kabul during the momentous period of the Soviet occupation of Afghanistan and three years as part of IPKF in Sri Lanka as also my visits to China and UK as part of Military delegations; I am convinced that there is no better soldier than the Indian Soldier. (Regrettably, the Col who wrote to Gen Kayani, the Chief of Pakistan Army has not had the experience of being in combat conditions with these gallant men). The Indian soldier can suffer the Privacy, his needs are few and more importantly, he can brave the hardships of Army life even in combat, better than any other soldier in the world. I therefore couldn't even fathom the very thought of his being under so much of stress that he takes the extreme step of either putting an end to his life or raising the hand that salutes his officer, to kill him.

It is in the quest of seeking an answer to the problem of Suicides and Fratricides in the Army, that I took up this study. Many senior officers said that there was nothing new in this phenomenon. Such incidents have been happening earlier and will continue to happen even in the future. In any case, so much has been written about it that another study would serve no useful purpose. They also retorted by adding that the Army and the MoD have done enough. So, what new was I to bring out?

My response to all those who harbour such thoughts is; if all has been done that was required to be done, then why are the cases of suicides and fratricides not on the decline? Infact, they are on the increase. The simple deduction is that either the Army has not done what it should have done or there is something more to be done. The basic question is that we cannot afford to lose valuable lives due to suicides and fratricides, as also let our men suffer the mental disorders that leave them to lead the rest of their life of indignity. It is in pursuance of reducing the human cost that I decided to take up this study.

My study begins with the reality that the officers and men of today are not the same as their predecessors, in many ways. They have been adversely impacted by the society from which they come and the exposure to TV/other Media. Their levels of aspirations are high and they feel that they have been left behind in the race of life, in comparision to their counterparts in the civil society. Their erstwhile recruiting base which was basically rural, has changed. There is an increasing number of Nuclear families, thus imposing greater burden of responsibility on them. Even the working environment has changed; in J&K the environment is intense and the rules of engagement are not the same as in the NE Region, imposing serious restrictions on them in conducting CI Operations. Stress, therefore is a byproduct of the environment from which they come and the environment in which they operate- the environment of work is hardly any different even in peace stations. On top of the unfavourable working environment is the domestic pressure on the soldier- he is caught between the two, which results in Stress. I, therefore see Stress as the mother of all other maladies.

The second issue that has been examined is the phenomenon of suicides, which is very complex and its causes are hard to find. Some individuals, by genes, are prone to depression and have suicidal tendencies. When such individuals face any problem, either they choose the escape route by resorting to Alcohol or go into Depression. When they realize that the problem is insurmountable or the disease is incurable, they tend to put an end to their life. The major cause of suicides is thus Personal/Domestic problems. This is as a result of some chemical transformation that takes place in their body. The study addresses this issue in detail and highlights the fact that, notwithstanding the fact that suicides is a world wide phenomenon and that our rates of suicides are less than the National Average as also the rates of suicides in other major armies of the world; however, we cannot take comfort in such statistics and need to address the issues that lead to suicides.

The third major issue that is addressed is that of fratricides. When an individual is provoked by another person either by hurting his IZZAT or by being insensitive to his sensitivities, the individual reacts violently leading to fratricide. The other major causes could be Injustice or Non-grant of leave in an emergency.

Besides these three major maladies, I have also examined other related issues like Alcoholism, Neurosis, Leadership, Motivation and Morale, Rashtriya Rifles, the tenures in LIC and women in the Army.

In my recommendations, I have taken the approach that the Army, being a national asset, cannot be left to fend for itself and there has to be a joint strategy to find a lasting solution to the problem. While the MOD should deal with the Systemic issues which are beyond the ambit of the Army, the Non- systemic issues must be dealt with by the Army. The book emphasizes on the need to reduce the pressure on the Army, improve the quality of life of the soldiers and the involvement of the State Government/Central Government officials to resolve the Domestic problems of the soldiers back at home.

I have tried to present the data as credible as it can be. The research is otherwise based on evaluation of several studies that had already being done. The reader might get an impression that the same studies have been referred to in all the chapters. But this is not so. Only the portions relevant to a particular chapter have been referred to, for evaluation, in that chapter.

Not withstanding the above, any errors or omissions on my part are unintentional and these, therefore may kindly be over looked. I would await suggestions from my readers.

Abbreviations

Ser	Abbreviation	Meaning
1	AR	Assam Rifles
2	Appx	Appendix
3	Avg	Average
4	Armr/Armd/AC	Armour/Armoured Corps
5	Arty	Artillery
6	AAD	Air Defence Artillery
7	ASC	Army Service Corps
8	AMC	Army Medical Corps
9	AOC	Army Ordinance Corps
10	BMP	Border Management Profile
11	Brig	Brigadier
12	CT	Counter Terrorist
13	CI	Counter Insurgency
14	Cas	Casualty(ies)
15	CPC	Central Pay Commission
16	CRPF	Central Reserve Police Force
17	CISF	Central Industrial Security Force
18	Col	Colonel
19	CC	Central Command
20	CPO	Central Police Organisation
21	CMP	Corps of Military Police
22	DIPR	Defence Institute of Psychological Research
23	DSC	Defence Security Corps
24	D Force	Delta Force

25	EME	Electronics and Mechanical Engineers
26	ERE	Extra Regimental Employment
27	EC	Emergency Commissioned
28	Engrs	Engineers
29	Edn	Education
30	Fd	Field
31	HAA	High Altitude Area
32	HR	Human Resources
33	IED	Improvised Explosive Devices
34	IS Force	Internal Security Force
35	ITBP	Indo-Tibetan Border Police
36	Inf	Infantry
37	IZZAT	Respect
38	JCOs	Junior Commissioned Officers
39	J&K	Jammu and Kashmir
40	Jawan	Urdu word for a soldier
41	Kilo Force	HQ RR= Kupwara
42	LIC	Low Intensity Conflict
43	LOC	Line of Control
44	Lt Col	Lieutenant Colonel
45	Lt Gen	Lieutenant General
46	Lve	Leave
47	MHA	Ministry of Home Affairs
48	Maj	Major
49	MOD	Ministry of Defence
50	MH	Military Hospital
51	NCO's	Non Commissioned Officers
52	NK	Not Known

53	NW	North West
54	NC	Northern Command
55	NDW	Negligent Discharge of Weapons
56	Offrs	Officers
57	OR's	Other Ranks
58	PTSD	Post Trauma Stress Disorder
59	Psy	Psychological / Psychology
60	PBOR	Personnel below Officer Rank
61	PMF	Para Military Force
62	Refer	Reference
63	RR	Rashtriya Rifles
64	RVC	Remount Veterinary Corps
65	Romeo	HQ RR-Rajouri Force
66	SC	Southern Command
67	SWC	South Western Command
68	Sigs	Corp of Signals
69	SII	Self Inflicted Injury
70	TD	Temporary Duty
71	V Force	HQ RR- Victor Force.(Valley)
72	WW I	World War I
73	WW II	World War II
74	WC	Western Command

CHAPTER 1 - PRELUDE

India won independence but failed to govern it. Lack of good governance has led to economic and social disparities. This social injustice and economic inequity has compelled sections of our society in different parts of the country, to take up arms to seek justice, leading to situations deteriorating well beyond normal law and problems necessitating the employment of the Army.

The Indian Army has thus been exposed to LIC/ Terrorism or Militancy since the 1960s. Since then, new areas of conflict have emerged every decade, with more and more troops getting involved. Beginning with the problem of Nagaland in 1960s, the insurgency spread through other parts of the NE Region to Mizoram and Assam in the 1970's, then to other parts of the country like Punjab in the 1980's and J&K in the 1990s.The Indian Army was also engaged as a part of the Indian Peace Keeping Force (IPKF) in Sri Lanka from 1987 to 1990 and later as part of the UN contingents in Somalia and other parts of the world. The Indian Army has thus been conducting CI Operations for longer than any other Army in the world. The hall mark of such an involvement has been that they have carried out these assignments successfully, without being defeated like the Americans in Vietnam and the Soviets in Afghanistan. This speaks highly of the tenacity of the Indian soldier and the professionalism of the Indian Army officers.

However, the fine performance has not been without a price; both in terms of human and financial cost. Such commitments have taken a heavy toll and presently the Army has been afflicted with the maladies of Stress, Suicides and Fratricides. This has led to **crises within the Army** as, more than those who were killed in encounters, many soldiers have been incapacitated due to Neurological or Psychiatric disorders.

The problem is that instead of reducing, the cases of Suicides and Fratricides are on the increase. It will thus be fair to assume that enough remedial measures have not been implemented at various levels to mitigate the problem and more needs to be done.

Several articles and books have been authored and studies have been carried out to address the problem. But most of the studies conducted so far have been by Psychologists or Psychiatrists and that too by middle level officers. No study has been instituted at a senior level by an officer with field experience, as was done in Vietnam (though the situation in the Indian Army is not like that of the US Army in Vietnam). These studies have also not defined the role of leadership in LIC environment and ignored some issues at the strategic level which contribute to stress in the Army. These studies have given useful inputs no doubt but as Dr Pestonjee has said, "What can be treated medically are the effects of stress and not its causes". This study has been carried out in the context of the aforesaid.

The aim of the study is to analyze the crises within the Army that have been caused by maladies like stress, suicides and fratricides with particular emphasis on LIC environment, its impact on the Army and suggest measures to mitigate the effects of these maladies.

Scope

The study is confined to ascertaining the causative factors of Stress, Suicides and Fratricides and suggests remedial measures for these maladies. The clinical part of these maladies is best left to the medical authorities to address. Accordingly ,the scope of the study is to focus on the following aspects:-

- To understand the phenomenon of Stress, Suicides and Fragging/ Fratricides.

- To analyze the different accentuating factors of stress like the socio-economic changes in the society and their impact on the Army.

- To study the nature of Low Intensity Conflict as it differs from the Conventional war and how it imposes a psychological weight on the military mind.

- Determine the various manifestations of stress and their impact on the Army with particular reference to suicides and Fratricides.

- Determine the Stressors particularly in Low Intensity Conflict.

- Determine the causative factors of Stress, Suicides and Fratricides in the Army

- Determine the various sensitivities of the soldiers and how the leadership should respond.

- Understanding the role of Leadership- Led relationship and Morale and Motivation. Determine whether the Leadership is sensitive and caring towards the soldiers or not?

- To determine the ideal tenure in LIC environment without adverse Psychological effects.

- Determine the contributory factors or Systemic issues which have a bearing on the stress in the Army.

- Determine the factors that assist in the Management of Stress.

- Analysis of the problem of Stress, Suicides and Fratricides.

- Recommending measures for mitigating Stress, Suicides and Fratricides in the Army.

- Suggest the way ahead.

The study has taken the data up to Dec 2008.

Methodology

The study has not followed any scientific method but adopted a simple approach of examining and analyzing the study material: available through Internet, other studies and publications. It is also based on the answers to the questionnaire sent to both serving and retired officers as well as interactive sessions with them. It draws upon the enormous personal experience in Counter Insurgency Operations. Most importantly it relies upon the factual data and statistics provided by credible sources.

Layout

The various chapters are as follows:-

- Chapter 1   -   Prelude

- Chapter 2   -   Stress and its Manifestations

- Chapter 3   -   Accentuating and Contributory Factors of Stress

- Chapter 4   -   Causative Factors of Stress

- Chapter 5   -   Suicides in the Army

- Chapter 6   -   Fratricides in the Army

- Chapter 7   -   The Human Cost- Impact on the Army

- Chapter 8   -   Leader- Led Relationship and Morale and Motivation in CI Operations

Conclusion

While the study draws its conclusions, based on the analyses of other Studies/Reports and interaction with the veterans, it also reflects the perceptions of the author who has had long experience of conducting or witnessing CI Operations, both within and outside the country.

CHAPTER 2- PHENOMENON OF STRESS AND ITS MANIFESTATIONS

'One Man's Stress Is Another Man's Challenge'

(Mc Grath)

General

According to D M Pestonjee, an eminent sociologist and management expert,[1] the twentieth century has been called the "Age of Stress". Stress is every where; in the work place, at home and even on the streets. Every one operates in an environment full of uncertainties, turbulence and even intolerance. Competition has increased manifold with attending consequences- both good as well as bad. Every body is in a hurry and is running. In this race to go ahead, stress levels have gone up. We find a larger number of younger people suffering from psychological distress, depression, anxiety, hypertension and physical ailments. Stress has therefore become an inevitable companion of all today in all walks of life. Armed Forces, by no means are an exception to it and are not insulated from it. In fact they are exposed to situations that are even more stressful.

This gives the mistaken impression that stress is a recent phenomenon, unknown or non existent in the past. This is not true. We get enough evidence dating back to a couple of thousand years in Eastern scriptures, wherein scriptures to cope with stress have been highlighted. Yogic literature, meditation techniques and

---

[1] Book on Stress and Coping by Dr D M Pestonjee

respiratory controls are some of the known evidences. It shows that stress has been a subject of exploration and study for a long time.

The second issue is that environment plays an important role in influencing day to day behaviour. Any environment can be perceived to be either facilitating or inhibiting. The stressful environment of insurgency areas either motivates a person or causes tension and frustration. These frustrations affect the mental peace and cause emotional disturbances, which have a direct bearing on the performance of duties. Physical aspects of the environmental realities may or may not be altered, modified or controlled. These aspects in a combat situation are rather inevitable. But the human mind to an extent, can be controlled and conditioned and thus the perception modified. It would mean that since the influence on behaviour is through the media of perception, it could be controlled. Resultantly, one can avoid or at least mitigate the negative effects on the performance. Managing stress therefore appears to be a real need. Higher stress levels, besides manifesting in poor performance are also instrumental in initiating inter personal violence, running amok, and suicide. These are not only detrimental to the performance of the soldiers but also tarnish the image of the Army. The military leaders and those connected with the affairs of the Army, must, therefore understand the various nuances of this phenomenon.

Every one faces different kinds of challenges at one time or the other from childhood to old age. A child finds it difficult to go to school and cope with the examinations. An adult finds it hard to start a career, change the job, get married and have children. Finally, every one faces the dilemma of retirement, ailing health and adjustment with children, especially when one of the life partners passes away. During this course of journey of life, a person faces several challenges. The reaction or response to these challenges is Stress. This chapter gives the theoretical background of Stress so that officers can comprehend its various dimensions and understand the behaviour of their men better.

Section 1 – Understanding of Stress

What is Stress? [2]

Loosely speaking, Stress is the reaction of our mind and body to any change. It creates physiological and psychological reactions, making the body generate more energy. Thus stress prepares the body and mind to meet any situation.

Stress is different things to different people. To a mountaineer, it is the challenge of pushing physical resources to the limit, to reach the mountain top. To the homebound motorist, it can be the hassle of heavy traffic and toxic exhaust fumes. To the student, it can be the pressure of the examination. Similarly, to the soldier it can be the capture of an objective or killing/ capturing maximum militants.

Other Definitions.[3] Some of the other definitions are: -

- Stress is a constraining or impelling force e.g. under stress of poverty.

- Stress is an effort or demand on energy e.g. when an individual is subjected to great stress.

- Stress is a force exerted on the body.

- The Penguin Medical Encyclopedia defines stress as any influence which disturbs the natural equilibrium of the body and includes within its reference, physical injury, exposure, deprivation, all kinds of diseases and emotional disturbances.

- Application of Strain, Pressure or Force on a system.

- Whole process by which we appraise and respond to events that challenge or threaten us.

---

[2] Hand book issued to the Army on Management of Stress - 1.

[3] Different Sources- Definitions of Stress

- Events and Conditions that put a strain on the organism and pose a challenge to adjustment.

- A fairly predicable arousal of Psycho- Biological, (Mind, Body) which, if prolonged, can fatigue or damage the system to the point of malfunctioning or disease.

- Stress is the reactions to Excessive Pressure or other types of demands placed on individuals. It arises when they worry that they cannot cope.

- Stress is the wear and tear that our inds and bodies experience, as an attempt to cope with our continually changing environment.

- S= PTR, Stress occurs when the pressure is greater than resource.

- Stress is like a fire- more like embers- permanent embers. We can control them to be harmless embers and enjoy their glow or fan them into huge fires. Not knowing the better, many of us do the latter.

- The non-specific response of the body to any demand made on it-'Style'.

- **Stress occurs when the pressures upon us exceed our own resources to cope with those pressures –'Tom Cox'**

Stress Kills. Its' fatal effect on the heart is well documented. But new research suggests that it is probably the reason for everything. Chronic stress has been linked to all ailments as diverse as internal problems, Common Cold, Gum diseases, Erectile Dysfunction, Diabetes, growth problems and even Cancer. [4]

---

[4] *Times of India, Delhi Times*, 12 Jan, 2008

Dr Pestonejee writes, "Stress is unique in the category of diseases. It has no biological carrier such as a germ or a virus. Rather it is the result of how our body and mind function and interact." It is Psychosomatic in the true sense of the word where *Psyche* means, mind and *soma* meaning body. It is a consequence of how we regulate, or to put it more appropriately, how we do not regulate the mental and physical functioning of our being. It is the disease created by the abuse of our minds, and bodies that lead to different symptoms to different people.

Misconceptions.[5]

- All Stress is harmful.

- Must be avoided.

- Hard work Kills.

- Not organisational problem.

Some Interesting Facts.[6]

- Disease of the 20th Century.

- Stress and Pollution Relationship.

- 70% of all patients in developed countries visit doctors due to stress diseases.

- 60% Medicines related to stress.

- $ 100 billion on stress related treatment (US- $ 40 bn).

- $ 150 billion spent by industry in changing Executives and Leave.

---

[5]　Hand Book on Management of Stress Part 11.

[6]　Same as 4 and 5

- UK, 3.5% GNP i.e. Defence budget spent on stress.

- Estimated that at any point of time, 10% population in India suffers from psychological disorders.

- In the Armed Forces too, IHD, BP-Psychological disorders are on the increase particularly in younger generation.

- Research conducted by TATA Institute of Social Sciences (Mumbai) on 30 Corporations, in 1999, concluded:

    - Moderate to High stress   - 80-70%

    - Depression Prone   - 30-40%

    - Serious Marital Problems   - 30%

- Reason- Rapidly changing economic conditions and heavy demands on life of individuals.

- 37% younger employees suffer from depression, of this 8% have suicidal tendency.

Stress Model.[7] A simple model of stress shown below defines it as a 'constraining' force acting on a person who is trying to cope with this force, exerts or strains himself and feels fatigued or distressed.

---

[7] Dissertation on Management of Stress by Brig (now Maj Gen) S K Bhardwaj

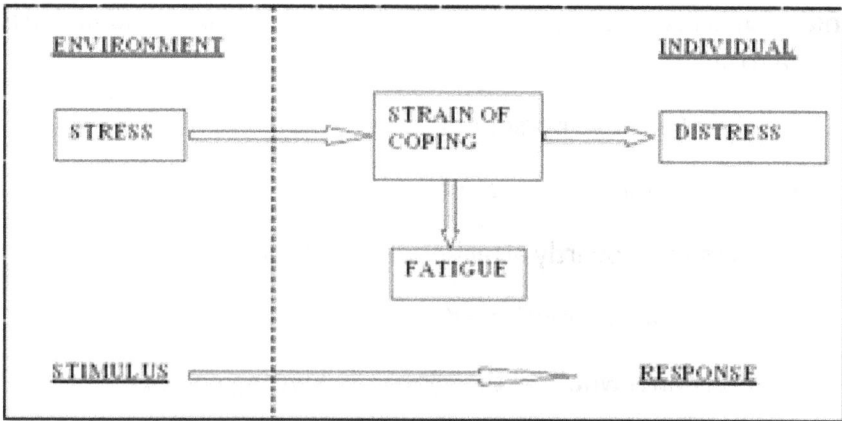

Layman's Model of Stress

Types of Stress

There are two types of stress:

- **Negative Stress**. It is a contributing factor in minor conditions such as headaches, digestive problems, skin complaints, insomnia, and ulcers. Excessive, prolonged and unrelieved stress can have a harmful effect on mental, physical and spiritual health.

- **Positive Stress.** Stress can also have a positive effect spurning motivation and awareness providing the stimulation to cope with challenging situations. This type of stress also provides the sense of urgency and alertness needed for survival when confronting threatening situations.

Other Aspects of Stress

Tolerance Level of Stress. It is an essential part of life. It is necessary to identify the optimum stress levels. If unattended to, it sooner or later manifests itself in the form of one or multiple diseases.

Individuals Prone to Stress.   The following types of personnel are prone to stress:-

- Over Sensitive personnel

- Over Ambitious personnel

- Personnel overtly attached to the family

- Personnel not motivated adequately

- Personnel who feel their job was a wrong choice

- Careerists

Classification of Stress

There is no straight method of classification. Stress can be measured in terms of degree of stress; High, optimum, severe, high optimum or low. It can also be expressed as transient or temporary or prolonged exposure. Stress can be Physical or Emotional depending upon its effect on the body.

Stages of Stress.

- *Honeymoon Stage*. A euphoric feeling of encounter with new job viz Excitement, Enthusiasm, Pride and Challenge.

- *Dysfunctional Features*. Energy resources gradually deplete- Habits and Strategies for coping with stress formed.

- *Fuel Shortage Stage*. Overdraw of reserves and the realization comes that the energy resources are depleted.

- *Chronic Symptoms Stage*. The Physiological systems become more pronounced and demand attention and help.

- *Crises Stage*. When these feelings and symptoms persist over a period of time, the individual enters this stage.

- ***Hitting the Wall Stage***. This is an experience that is so devastating that can completely knock off a person. With all the adaptable energy depleted, one may lose control over one's life and it may be an end of one's professional career.

- **Burn Out Stress Syndrome (BOSS)**.

- **Rust Out Stress Syndrome (ROSS)**.   This is opposite of BOSS and is indicative of Stress under load. It occurs when there is a gap between what the individual is capable of doing and what he is required to do. The concept of role erosion is close to the concept of ROSS. ROSS can arise both through qualitative and quantitative aspects of work.

- Any emotion is known to exert strong influence on judgment that relates to that particular emotion. To substantiate this, tests conducted to examine risk perception among fearful and angry individuals duly revealed that individuals made pessimistic risk preferences while angry people made optimistic decisions.

Section 2- Soldier and Stress

Violence, by itself, has become synonymous with trauma and stress. That is why over 80,000 people from Kashmir alone complained of depression during 2005-06 with over three fourth of them being diagnosed with serious psychological disorders. The people especially the children were also known to have suffered such psychological disorders in the War in Vietnam and Afghanistan.

Axiomatically, warfare, without doubt, is among the most distressing circumstances that human beings endure. To be separated from the family, friends and familiar locations for long periods is sufficient enough to produce stress reaction in most healthy people. Further, mere deaths and injuries to friends and colleagues are emotionally devastating and cause intense grief reactions. Add to that the constant threat of personal maiming or death and contributions to killing of others. The result is extremely high levels of psychological arousal, aggression, anxiety or fear. When stress is sustained for long time or if it is extreme and especially when both conditions of duration and intensity are combined, stress produces a wide range of negative effects that are not easy to counter act. Even the most mentally healthy and stable individual's experience some changes after exposure to operational/ combat stress. This is reflected in the letter written by Lt Gen Sir Thomas Picton, commander, British 5th Division to Lord Wellington in the battle of Waterloo, 1815, saying, [8]

> **"My Lord, I must give up. I am grown so nervous that when there is any service to be done, it works upon my mind so that it is impossible for me to sleep at night. I cannot possibly stand it and I am forced to retire".**

---

[8]   *Psychiatric Times-* Hidden Combat Wounds-Extensive, Deadly, Costly-10 Jan 06

Therefore from generals to the enlisted in the lower ranks, every one in the military is vulnerable to combat stress. Thousands of cases of officers and men were evacuated as cases of shell shock in the British forces. During World War I, 80,000 soldiers – one seventh of the disability discharges were due to shell shock. Later in World War II and Korean War, the US Army had to send home nearly 40,000 soldiers for Psychiatric treatment. It was during the Vietnam War that the Americans classified as Psychiatric cases. In 1973, almost 1/3 of Israeli and Egyptian casualties were cases of Psychiatric disorder. In more recent operations, the military has aggressively treated potential stress cases from Kosovo and Bosnia. Presently, the US and the British forces are experiencing Psychiatric disorders in Iraq and Afghanistan. The US Forces suicide rate is 19 per 100,000 annually. By one estimate, the number of Psychiatric cases in every war in the 20th century has exceeded the number of soldiers killed by hostile fire by 100 percent.

Stress Casualties

Some of figures for various Wars/ conflicts are as under:-

- Normandy   - 10%

- Korean War   - 3.7% of total strength.

- Vietnam War   - 1.5% of total strength.

- Yom Kippur War   - 60% of total casualties

According to another study on "Evolving Medical Strategies for LIC Conflict" by Lt Col Ajay Dheer et al, the Psychological casualties in various theatres of war are given in Table 1 below :-

Table-1, Psychological casualties in various theatres of war[9]

Theatre	Psy Cas%	PTSD%
WW II (Far East)	82	28
WW II (NW Europe)	80	11
Korea	73	47
Vietnam	-	35.5
Northern Ireland	50 suicides per unit in 6 months tenure	40 per year
Falklands	NK	22
Kashmir incl BSF	15*	7

* This figure appears doubtful

The Indian soldiers being a victim of combat stress should therefore, not come as a surprise. The environment of CI operations is no different than what the Americans and the British are facing in Iraq and Afghanistan. While one cannot take solace under the pretext that the rates of Suicides and Fratricide and other Psychiatric cases are less than that of the US and the British Forces, there is no need for an alarm. But it is definitely a matter of concern and the problem therefore needs to be addressed. The future is going to witness more and more stress related casualties as the future battle field environment is going to be characterised by increasing kill probability of weapons, improvements in night fighting capability and so on. As regards the effects of stress, Kulgan has said in 1976:

---

[9]   Evolving Medical Strategies for LIC by Lt Col Ajay Dheer et al

"People who fight future wars may experience so much strain that they will breakdown, whether they will come in contact with the enemy or not"

Section 3- Stressors and Response

Types of Stressors

Basically, there are two types of stressors:-

- External stressors.

- Internal stressors.

External Stressors.

- Physical Environment. (Noise, Heat, Bright Light, and confined spaces).

- Social Interaction. (Rudeness, Bossiness, Bullying, Aggressiveness by others).

- Organizational. (Rules, Regulations, Red Tape and Dead Lines).

- Major Life Events. (Birth, Death, Promotion, Job Change, Marital Status Change).

- Daily Hassles. (Commuting, Misplaced Keys, Mechanical Breakdowns).

Internal Stressors.

- Life style choices, (Caffeine, Lack of Sleep, Overload schedule).

- Negative self talk. (Pessimistic Thinking, Self Criticism, Over Analysing).

- Mind traps. (Unrealistic Expectations, Taking things personally, Exaggeration, Rigid Thinking).

- Personality traits. (Perfectionists, Workaholics).

Major Stressors Affecting Service Personnel. Refer to **Appendix A** attached.[10]

Life Stress Events. Refer to **Appendix B** attached.[11]

Stressful Events and conditions Experienced most often. In his report on "Medical Strategies for LIC" Lt Col Ajay Dheer et al has given a list of stressful events that generally occur in an LIC environment and these act as Stressors.

Stressful Events	Weightage %
No information	85
Separation from family	75
Monotony and Boredom	70
Uncertain Results	70
Indifferent Habitat Conditions	65
Fear of becoming a casualty	60

Response to Stressors

The responses or the reactions of these soldiers to this stress experience are of two types:-

- The Normal Response.

- The Abnormal Response.

---

10 and 11- Same as 2

Normal Response [12]

When the soldiers react to stress with an emotional or somatic (Physical) response, it is termed as Normal Response and does not constitute conditions of mental disorders. This response will also vary from individual to individual, depending upon the personality variables like childhood experiences, family history and the coping mechanism, developed as a result of these factors.

According to the Medical Appendix to Acute Stress Reactions including Acute Stress Disorders, Shell shock and Combat fatigue, there are three components to the Normal Response:-

- Fear, Anxiety and Depression. The emotional response to danger or threat causes fear and anxiety while the response to loss is sadness or a feeling of depression.

- Fatigue and other symptoms. The somatic (Physical) reactions to threat result in automatic arousal and include rapid pulse, dry mouth, increased muscle tension and Gastro-Intestinal sensations. This results in reduced Energy and physical activity.

- Psychological Mechanism. The third component of the response comprises of those Psychological Disorders that reduce the impact of the emotions endangered. In this the soldier unconsciously develops a variety of defence mechanisms to differing degrees in order to help him deal with the mental conflicts. Some of the most common psychological processes are:-

- Denial. Refusing to accept what some one would normally be expected to believe (Carries on as if unaware of a particular diagnosis).

---

[12] The Texas Blue- No Magic Bullet-Stress Programme by Dr Jeff Mitchell Http/ /www.The Texas blue.com

- Regression. Adapting to an earlier stage of development. (Passively being taken care of like a child).

- Displacement. The emotion is shifted to another less sensitive object (Being angry with colleague instead of superiors).

- Repression. Uncomfortable memories or emotions are kept out of awareness.

- Identification. Incorporating in one's own behaviour the characteristics of another to reduce the sense of loss.

- Other maladaptive strategies include actual avoidance or coming to terms with the situation. (This involves working through the unpleasant idea and emotion by recalling the event and finding a way of accepting that it happened). Such strategies include excessive use of Alcohol or drugs, aggressive behaviour and deliberate harm through self inflicted injury.

The Abnormal Response

Most of the behaviours fall within the normal range and most of the neurotic disorders are on a continuum with the Normal. However, if the stressor is sufficiently severe and/ or the soldier is vulnerable, various mental disorders develop. The stage at which symptoms constitute such a disorder or disease depends on several factors including severity and duration and other social factors. The disorders most commonly associated with exposure to a stressful event include Acute Stress Reaction, Post Traumatic Stress Disorder (PTSD), the adjustment disorders like phobia and other anxiety Disorders, Acute and Psychotic Disorders.

Acute Stress Reaction [13]

Clinically, Acute Stress Reaction is the transient response which sometimes occurs immediately following exposure to or during an exceptionally severe event which subsides within a short period of time (usually hours or days). Such a condition may be labelled as Combat Fatigue or Shell Shock. If the condition lasts longer than a month, then it is appropriately diagnosed as PTSD or an adjustment disorder.

The symptoms vary but generally include the following:-

- A feeling of detachment, disorientation, inability to concentrate or respond sensibly and being in a state of daze.

- Feeling of Arousal i.e palpitation, pounding heart, increased heart rate, sweating, trembling, shaking and dry mouth.

- Difficulty in breathing, choking, chest pain/discomfort, nausea, and churning in the stomach.

- Feeling dizzy, unsteady or faint, fear of losing control, or passing out.

- Muscle tension, Inability to relax, Feeling on edge or tense, a lump in throat or difficulty in swallowing.

- The Mind going blank, being startled, exaggerated response to minor surprise.

- Person appears to be out of contact with others but is not conscious or asleep.

- Following the trauma, the event may constantly intrude into awareness with extreme clarity. Amnesia of varying degree for the event may occur.

---

[13] Acute Stress Reaction- Medical Appendix

The diagnosis to acute reaction is only made in those cases where no pre existing mental disorder is evident prior to the episode, except for personality disorders. It therefore does not cover exacerbations of established mental disorders. The person must be exposed to an exceptional mental or physical stressor and the reaction usually occurs within an hour of the exposure to that stressful event. If the Stressor is transient, then the condition should begin to subside within 8 hours. If the Stressor is ongoing then the symptoms should subside within 48 hours.

The stressor must be of an extremely distressing event either mental or physical. The trauma may be experienced alone-e.g rape or assault- or in the company of groups of people e.g military combat. Stressors producing this disorder include natural disasters like floods, earthquakes, accidental man made disasters (serious road accidents with multiple or critical physical injury, airplane crashes, large fires) or deliberate man made disasters like bombing and torture in prisoner camps. The most important factors in determining whether a person will develop an Acute Stress Reaction are proximity to the event, the severity of the event and the duration of the exposure.

The condition may precede PTSD. However, the main differences between this condition and PTSD disorder are as follows:-

- PTSD arises as a delayed response to a stressful event. It often manifests itself only after a few weeks and can only be diagnosed after the symptoms have been present for a month. An Acute Stress Reaction passes within a few days.

- The symptoms of PTSD are very specific.

If Acute Stress Reaction lasts for more than a month, then the diagnosis should be changed to the most appropriate disorder including reaction to severe stress unspecified, the adjustment disorders, PTSD or the transient psychotic disorders, depending on the psychopathology present.

Difference Between Normal Response and Acute Stress Reaction

The main differences are:-

- Acute Stress Reaction is a severe reaction to an exceptionally distressing event. It disappears over a relatively short period of time. The duration depends, to some extent, on the duration of the exposure to the stressor.

- The difference between this disorder and the Normal Response is one mainly of degree; Acute Stress Reaction being more severe in its symptoms. The Normal Response to being in a stressful situation would include the sensation of the heart racing, the mouth being dry, trembling, emotional sweating (mainly axillary, palmar and planter), rapid breathing and a sense of fear.

It may be of interest to the reader to reflect and see if he /she ever came under any of the above conditions.

Section 4- Manifestations of Stress [14]

Having covered the clinical aspects of the Stressors and their reactions, it is now proposed to discuss the manifestations of Stress, which can be classified under the following:-

- **Anxiety Disorders**. Anxiety is defined as a sense of apprehension that shares many of the symptoms such as fear but builds more slowly and lasts longer. Though everyone does feel anxiety from time to time, a clinical disorder is one that demands a visit to a professional. This thrives in a situation like our troops were in –"Operation Parakram" where the wait for the enemy never seems to end. Unfortunately the limit of tolerance is limited even for a trained man and it is attainable. The cause for anxiety need not be the wait for

---

[14] Adverse Psychological Effects of LIC by Maj P Badrinath-*USI Journal-2004.*

the enemy alone: it could be any personal or professional reason. The symptoms of such a disorder include restlessness, difficulty in concentrating, insomnia, irritability, fatigue and muscle tension.

- **Depression Disorders**. Depression is a state where the affected person reaches a state of mental sluggishness coupled with repeated thoughts reliving a certain incident which brought him sorrow. One of the major causes of depression is the inability to help a certain situation like the ones our soldiers encounter when they are deployed for CI operations. Operation Parakram was another situation where the soldier may have felt his inability to help the situation. Depression can manifest itself as a lack of interest, lack of alertness or even violent incidents due to prolonged spell of depression.

- **Acts of Indiscipline**. A soldier like any other human being also seeks to vent his frustrations in difficult times. This is true when he is affected by a condition of anxiety and depression. The incidence of over staying leave, desertion or for that matter even "Fragging" can be clubbed under this heading as they are not considered normal reactions of a trained soldier. While the incidents relating to leave can be dealt with, in the conventional manner as is done in the Army, incidents of fragging need to be examined in greater detail. There have been cases where soldiers who ran amok had to be let off after the trial, since they were not in a right mental state when the act was committed. However, plain trigger happy soldiers had to be brought to book to ensure that there is no recurrence of such incidents.

- **Suicide Tendencies**. Suicidal tendencies are commonly known to occur in persons suffering from depression, as is evident from the experience of daily life. A loaded weapon at the disposal of a soldier is fraught with danger in such cases.

- **Specific Phobia**. Specific phobias are basically overpowering fear of a specific object or situation, often accompanied by extreme anxiety situations. A possible cause could be the fear of an impending suicide attack. Though these types of phobias do occur and disappear normally, there can be problems if the duration of exposure to such incidents is high.

- **Post Traumatic Stress Disorder (PTSD).** A repeated, arduous reliving of a horrific incident over an extended period of time, leads to PTSD. In this condition, the affected person may display symptoms of anxiety, depression, phobia and suicidal tendencies. Approximately 22,400 Americans suffered from PTSD post 9/11. Going strictly by numbers, there could be quiet a few of our men who would also be suffering from this malady, but we can probably breathe easy, for our soldiers are much more resilient. However, we cannot be complacent about it.

- **Psychosomatic Disorders**. As per the dictionary meaning, Psychosomatic is a disease, physical in origin, both of mind and body. The list of psychosomatic symptoms which occur primarily because of stressful feeling includes stomach problems, heart irregularities, breathing difficulties, fatigue, headaches, muscle tension , skin problems, and elimination difficulties. Beside these common symptoms, others include hypertension, ulcers, allergy, asthma attacks and insomnia.

- **Impaired Decision Making**. It is well known that emotions in general do affect one's decision making ability and judgement. There is also a more intimate connection between certain stressful emotions like fear and anxiety and the process of judgement. It is found that dispositional anxious individuals rely on their anxiety to inform certain subsequent judgement.

Conclusion

This theoretical background has been provided basically for our Officers who are engaged In CI Operations. It will help them to read the symptoms and determine the malady from which a soldier is suffering before they manifest into a case of indiscipline or a case of Suicide or Fratricide.

Appendix A
(Refers to Chapter 2, Page 20)

MAJOR STRESSORS AFFECTING SERVICE PERSONNEL		
Personal	**Financial**	**Relational**
Value conflicts	Financial Aspirations	Break up of Joint Family system
Bereavement	Financial status vis-à-vis A civilian.	Separation from Family. Isolation from accustomed source of affection.
Frustration	Materialistic outlook	Health problems of the family
Competition	Deprivation of social concomitant	Deprivation of social and concomitant social satisfaction.
Health	Property disputes and Litigation	Property dispute and Litigation
Physical Discomfort	Superannuation and Alternate Job	Loss of Comrades
Restrictions of move		
Enemy Fire		
Lack of Privacy		
Sense of not being counted as an Individual	Occupational (Routine)	Occupational Stressors (Combat)

contd/-

MAJOR STRESSORS AFFECTING SERVICE PERSONNEL		
Personal	**Financial**	**Relational**
	Monotony, Repetitive and similar Tasks	Threat to Life, Limb and Honour
	Job and Task Anomaly	Loss of Comrades
	Round the clock avlb	Sound of wounded and dying men
	Over Load of Work	Long periods of enforced Boredom and anxiety between actions
	Under Utilization of Abilities	Intense and sustained Operations
	Role Stress	Fear, Tension and Anxiety
	Supercession	
	Strict Discipline and Regimentation	
	Frequent Posting and Moves	
	Pressure from Superiors	

Appendix B
(Refers to Chapter 2, Page 20)

LIFE EVENTS STRESS-(SOCIAL READJUSMENT SCALE)

Ser No	Life Event	Mean Stress Score
1	Death of a Spouse	95
2	Extra Marital Relations of Spouse	80
3	Marital Separation / Divorce	77
4	Dismissal from Job	76
5	Detention in Jail- Self/ Family	72
6	Lack of Child	67
7	Death of Family Member	66
8	Marital Conflict	64
9	Property/ Crops damaged	61
10	Death of a Friend	60
11	Robbery/ Theft	59
12	Excessive Alcohol/Drug Abuse	58
13	Conflict with In Laws	57
14	Broken Love/Engagement	57
15	Major Illness/ Injury	56
16	Son/Daughter Leaving Home	55
17	Financial Loss	54
18	Illness of a Family Member	52
19	Trouble Working with Colleague / Subordinates	52
20	Prophecy- Astrologer.	52

contd/-

Ser No	Life Event	Mean Stress Score
21	Wife's Pregnancy	52
22	Conflict Over Dowry	51
23	Sexual Problems	51
24	Self/ Family Member Unemployed	51
25	Lack of Son	51
26	Large Loan	49
27	Marriage of Sister/ Daughter	49
28	Minor Violation of Law	48
29	Family Conflict.	47
30	Break up with Friend	47
31	Purchase/construction- House-	46
32	Death of Pet	44
33	Failure in Exam	43
34	Appearing for Exam/ Interview	43
35	Getting Married/ Engaged	43
36	Trouble with Neighbour	40
37	Unfulfilled Commitments	40
38	Change of Residence	39
39	Change in Business	37
40	Outstanding Achievement	37
41	Begin/ end Schooling	36
42	Retirement	35
43	Change of Working Conditions	33
44	Change in sleeping Habits	33
45	Birth of Daughter	33

Ser No	Life Event	Mean Stress Score
46	Addition-Family Member	30
47	Reduced Family Functions	30
48	Change in Social Activities	28
49	Change of/in Eating Habits	28
50	Wife Starts Work	25
51	Going on Pleasure Trip	20

CHAPTER 3- ACCENTUATING AND CONTRIBUTORY FACTORS OF STRESS

There are two factors that may indirectly be the causes of Stress; these could be termed as Accentuating Factors and the Contributory Factors. These are discussed in subsequent paragraphs.

Section1- Accentuating Factors of Stress[15]

The changing times through which the Army has traversed during the past over six decades, the Politico- Bureaucratic establishment, the socio-economic changes and consequent changes in the values of the society and the advent of a booming electronic media and telecommunication revolution have had their direct impact on the man in the uniform. These factors have accentuated the problem of Stress, Suicides and Fratricides if not created them. It is therefore very important to understand these factors before addressing the issue of the "Causative Factors of Stress".

The Changing Role of the Army and its Impact

Traditionally, the Indian soldier came from the rural background. Though less educated, he was hardy and could easily brave the hardships of Army life. In yester years, Stress was never heard of. The Indian military history is replete with incidents and descriptions of steadfastness in the face of seemingly insurmountable adversity. Under the British era, the Army was the most coveted and respected

---

[15] Management of Stress in the Army by Maj Shailender Singh Arya-*USI Journal, 2010.*

organisation. The men in uniform were held in high esteem by the society. Any petition by a soldier was, therefore addressed with top priority. Even after independence, the ICS officers were directed towards the welfare of the soldier.

Later, during the 1950's and 60's, the soldier was still viewed as the saviour of the nation. But the humiliating defeat during the Sino-Indian war of 1962, gave a setback to the fine image of the Army. But soon thereafter, it revived after the 1965 war with Pakistan. The humiliating defeat of the Pakistan Army in 1971 gave further fillip to the image of the Army and the uniformed man continued to enjoy high status in the society.

The 80s and 90s were the turbulent decades for the Indian Army. Beginning with the militancy in Punjab in the mid 80s, soon an uninterrupted cycle began in 1987, with the innocuous induction of the IPKF in Sri Lanka. It was soon involved in a war, which was neither intended nor it was structured to fight. As it was being recalled even before completing its mandate, insurgency erupted in the valley in the 90s and later graduated to a full proxy war. Since then the Army has been engaged in two widely separated regions, fighting proxy war in J&K in the West and insurgency in the NE region. In 1999, there was a brief focus on fighting conventional war in Kargil. It was the first war which got televised for the nation and the world. Though it briefly revived the status of the soldier as the national hero but the euphoria of victory was short lived. This brief interlude did not alter the harsh reality of the Army remaining engaged in CI operations Since in this type of conflict, there are no visible signs of victory, the soldier was no longer considered the saviour of the nation thus losing his status (IZZAT).

Socio- Economic Changes and consequent changes of the values in the Society [16]

In 1991, Dr Manmohan Singh, the then finance Minister of India, began the process of liberalising the economy. It soon paid rich dividends by the turn of the 21st century and the Indian landscape changed in an unprecedented manner. The impact was there to see.

Today, the nation, after long years of slumber and a sluggish economy has broken free from shackles of a bureaucratic control and the Inspector Raj. The restrictive policies of control have given way to economic liberalisation. The nation has moved into the orbit of globalisation and emerged as a developing nation or a nation under transformation. This has ushered in an era of a new culture of consumerism as the dominant trend. In today's environment, therefore the traditional values are undergoing a marked transition. Simplicity has been replaced by materialism, modesty has given way to aggression, patience has been replaced by a tearing hurry and above all, ambition has replaced contentment. The people have been impatient and the desire to have the latest consumer goods has transcended all boundaries of urban, rural, rich and poor divide. This has led to excessive spending on acquiring luxury goods and a house / apartment at a young age against loan. Savings have taken a back seat. The urge to earn more has led to urban migration. Caution and value are being flouted. Inter caste marriages and subsequent divorces are on the increase. Traits like greed, cleverness, acquisitiveness and aggressiveness are admired traits while qualities of kindness, generosity and understanding are seen as the concomitants of failure. Seeking jobs in corporate sectors / foreign firms' attendant with comfortable surroundings and lesser risks but high earnings

---

[16] Changing Socio- Economic Values and their impact on the Armed Forces by Maj Shailender Singh -*USI Journal, Apr-Jun 2007.*

are the preferred choice. Government jobs, particularly a career in the Armed Forces appears to be the last option. This transition is inconsistent with the core values of the Armed Forces, depriving them of the best material in the youth and resulting in the continued shortages of officers in the junior ranks.

The socio-economic changes in the society and consequent changes in values have impacted the life of our people. Nuclear families are now the norm rather than an exception. The land holdings have been reduced to a pittance in rural areas. Seeking jobs for both men and woman has become a compulsion. The working woman no longer enjoys the traditional support of the joint or an extended family, resulting in a plethora of functional problems in sustenance of such nuclear families, particularly of those in uniform. Because of exigencies of service, it poses difficulties for such uniformed men to strike a balance between their homes and the professional front. The management of separation has led to an intense experience for many inexperienced families with the potential to cause marital discord

In essence, therefore, the officers and men of today come from a society where there is a decline in the values- principles and standards. The temptation to bend is strong. Many succumb to pressures. They justify their actions by saying-'every one is doing it'. They make excuses and console themselves saying; values have changed. Society appears helpless in the face of the onslaught on its values. As individuals they are aggressive, impatient, intolerant and of being materialistic from child to the middle aged. Yet the same society has high expectations from its Armed Forces. It sees the Armed Forces as the last bastion of virtue. It fervently wants to believe that the Armed Forces are different. Even the Armed Forces themselves view themselves in the same light. This is a flattering thought but it is not realistic.

Core Value Deficiet

The officer in the Army must be different from those in the civil services. The reasons for being different are not far to seek. In the Army, the stakes are high. The officer leads his men into battle and he expects them to follow him in the most adverse circumstances even to the point of death. The Armed Forces have well established traditional core values of selfless devotion to duty, honour, honesty, integrity and loyalty along with moral and physical courage. These values are absolutely non-negotiable. It is the unflinching devotion to duty, unquestionable loyalty and very strong sense of honour (Izzat) of the Regiment (Paltan) that has led to countless men making supreme sacrifices of life in the service of the nation. These core values have motivated and bonded together every man in the services into well integrated and cohesive units / sub units. These men stand erect on the pillars of physical and moral courage. The other unique characteristic of the Armed Forces is that they are apolitical and comparatively corruption free.

However, the Armed Forces, being a part of the society could not remain isolated from the environment of scams and scandals; corruption came to be valued against Honesty. The lure of making quick money trapped many greedy officers. Some of them, especially in the higher ranks became privilege seeking, thus destroying impeccable reputations, built over a long time. This new value system led to an increase in incidents of petty crimes and scams, not in keeping with the core values of the Armed Forces.

Altered Recruitment Pattern

The socio- economic changes have impacted not only the recruitment pattern of the men but also the intake of officers. The basic qualifications for various ranks have been enhanced with the minimum being a Matriculate. But there are also a large number of XII standard and Graduates. The proportion of recruits from the rural areas has reduced with corresponding increase in favour of the urban based. The average recruit of today is educated, aware and ambitious with high levels of expectations/ aspirations. The innocent looking, hardy and frugal soldier is a thing of the past. Similarly, while earlier the recruitment was from agricultural background, now days it is drawn from different backgrounds. The officer cadre is increasingly being drawn from lower/middle class families. Some of them have probably poorer social background than the men from the same village. Thus the gap between the leader and the follower has narrowed further. However, the officer of today is better educated and professionally more sound but there are some areas of disquiet.

Perceived Neglect and Exclusion

Since the Third Pay Commission, the Armed Forces personnel have perceived to have been neglected vis-à-vis their counterparts especially the IAS/IPS cadres. The subsequent Pay Commissions have only further confirmed the neglect and disregard towards the men in uniform. This was articulated by the veterans after the 6th Pay Commission through depositing of Medals with the President of India, News room discussions and TV debates. To- day the Armed Forces feel that they have little or no say even in matters that directly affect them. The soldier therefore not only lost his voice in the affairs of the state but also the place of pride, that he had in the society.

The other development was the new power equation between the Politician and the Bureaucrat. They became the power centres. In such a power equation, the soldier became insignificant for the police and the district administration –that could now be influenced mainly by money or political power. Since the soldier had none of these, his petitions and problems got relegated to a lower priority.

Active Media and Public Awareness

The Armed Forces have been surprised with two developments; Active Media and Public Inquisitiveness. The Media till recently had not interfered with the affairs of the Armed Forces. But the indulgence of the Media in sensational reporting has brought the Armed Forces under their watchful eyes. This has resulted in the media covering events which otherwise were considered sensitive by the Armed Forces; the latest example being the coverage of the Kargil War in context to the hearing of Brig Devinder Singh and the decision by the Armed Forces Tribunal.

Advent of Modern Technology

There were two developments that transformed the lives of the society and impacted the soldier; TV and Telecommunications. Access to TV has increased awareness about contemporary issues such as public opinion, rights and privileges, evolving life styles and expectations arising from it. It impacted the life styles of the men in uniform and raised their hopes and aspirations. Gadgets like the PCO, Internet and the Mobile transformed the way people communicate. The possession of the Mobile enabled the soldier to remain in constant touch with his family. This had a positive as well as a negative impact. While good news from the family was a motivational factor, bad news could lead to Anxiety and Depression. It thus acted as a double edged sword. A fine management act in the Management of the TV and the Mobile in particular was necessary so that these contribute positively.

Impact

What has been discussed above has seriously impacted the Armed Forces personnel. Some of the major issues are highlighted as under:-

- The major impact has been that the events of the past over 60 years have made a significant change in the status of the soldier and therefore his respect in the society. He was no longer the saviou r or the Hero of the nation. A new set of people like the cricketers, cinema heroes, the neo rich, the bureaucrat (Babu) and the Politician (Neta) took the centre stage. The soldier got reduced to an object of sympathy, This is the reason why hardly any action is taken either by the centre or the state officials to address the problems being faced by servicemen at home.

- Similarly, with the amendments in the "Warrant of Precedence" the soldier moved downwards as compared to the IAS and the IPS cadres. This had a corresponding adverse effect on his pay and allowances as well as the IZZAT in the society.

- Due to the Armed Forces losing its élan, the young generation did not find this a preferred career choice resulting in acute shortage of officers especially at the junior level. This shortage adversely affects the operational performance of the units.

- Not withstanding the fact that the Army has preserved most of its values intact, certain negative values having permeated into the men in the services, cannot be overlooked.

- The TV and other media exposed the soldier to what was happening outside of the world he was living in. His aspirations and expectations were aroused and he felt that he was being left out, compared to his colleagues in the home town who made easy money.

- The break up of the joint family system burdened the soldier with greater responsibilities towards his family especially the wife and children.

- The advent of the mobile phone gave him easy access to his family. While giving him the advantage of keeping in touch with them regarding their wellbeing, it also got him involved with trivial issues, thereby disturbing his peace of mind.

- With the new type of recruits from the urban areas also enlisting themselves, the Army leadership had to manage two different types of men, thus posing problems of Man Management.

It will be fair to assume that while the nation has surged forward to being one of the major powers of the world; the advent of new technologies, adoption of bold initiatives having led to liberalising of the economy and the emergence of the media as a powerful weapon to mould opinions, have changed the lives of the people. More than anything else, these events have produced a new type of society; with values very different from those of their predecessors, resulting in the emergence of a new brand of heroes. The major impact has been felt by the Armed Forces wherein the uniformed man has been relegated to an insignificant position, with loss of his place of pride and IZZAT in the society. His welfare and well being is being neglected with no cognizance being taken of his requests. That is at the base of all the maladies that the soldiers are suffering from with different manifestations. This is best reflected in the following quote[17] :-

"When a Military Spirit forsakes people, the profession immediately ceases to be held in honour and Military men fall to the lowest rank of public servants; they are little esteemed and no longer understood——Hence arises a circle of cause and consequence from which it is difficult to escape-the best part of the nation shuns the Military profession because that profession is not honoured and the profession is not honoured because the best part of the nation has ceased to follow it."

- Alexis de Tocqueville on Peace Time Armies

---

[17] Challenges for Military Leaders of Future due to changing Socio- economic Norms by Lieutenant Commander Yogesh Athavale-USI Journal,Jan-Mar 2010

Changes in the socio-economic milieu in the country, which directly impinge on the quality of domestic life of the soldier, have hardly helped matters, in this regard. With the security of the joint family system having gone; with the rising expectations of all individuals for a better quality of life; with the accentuation of problems related to land, housing, education and career development of the children; with increasing peer pressure to vault to a higher threshold of materialistic life and with the increased level of awareness spurred on by the media, the soldier today can hardly be expected to be a blind slave to the straight-jacket of military discipline, without analysing its wider implications.

However, there is nothing wrong in the above. Changes in the outside environment would surely manifest itself in the Armed Forces as General Sir John Hackett in the Profession of Arms has said, *"It has frequently been emphasized in what has already been said that the pattern of parent society is faithfully reflected in the military institutions to which it gives birth and that developments in the first are followed sooner or later by corresponding developments in the second"*. The officers have to understand the social and institutional changes and its effects on soldiers, real and important problems of soldiers and take appropriate action.

Section 2- Contributory Factors of Stress

Stress experienced by an individual is a product of two factors;

Basic Anxiety and Environment. Stress due to Environment is job related but the phenomenon of Basic Anxiety is rather complex.[18]

Basic Anxiety. This is based on the experience of :-

- ***Birth Trauma.*** In the pre natal stage of development when

---

[18] Extracts of Dissertation on Management of Stress in the Army. Same as 7.

the child is in its mother's womb, it is a blissful state. But when the child is born, its contact with the blissful state is broken and it feels many pressures like those of gravity, noise, light, need for oxygen etc. It adjusts to the new environment gradually. But it appears that certain amount of feelings related to birth trauma never leave the man and is carried by him to the years of adulthood also. Therefore, **four fears; fear of darkness, fear of failing, fear of getting suffocated and fear of the unknown are part of human nature or instinct.**

• *Dependency Syndrome*. During the first five or seven years of development, a child is entirely dependent upon others for survival. Not only are its requirements of food, clothing and shelter etc met by them, but they are also responsible for building up its confidence to combat the above mentioned four fears. This prolonged dependency upon others during the formative years instills a sense of inadequacy and inferiority in a man. There is nothing wrong so long as it remains within limits. It is the degree of anxiety which makes different people react differently to the same or similar situation of stress- some getting alerted by little stress, others taking much of it in the stride. In day to day parlance, we call the former as worrying type and the latter as those who couldn't care less.

Environment. The second important factor is the environment in which the soldier works and the environment he comes from; that is the society. The factor of society and its impact has already been discussed earlier. In this chapter, it is proposed to discuss the environment in which the soldier works. It is a well known fact that the Indian soldier in deployed in varied environments. But broadly, these are classified into two categories- Field and Peace Stations. While the connotation of Peace stations is simple, the Field area has different terrain conditions, varying from the Glaciers of Siachin to the snow clad or the forested/barren heights of J&K as well as the NE region and the Plains Sector Operationally, the troops in the Field area are deployed either in the Border Management along the LC or the LAC and CI/CT operations. This study has focussed on the conditions prevailing in LIC environment and the Peace station.

Environment in LIC

Before describing the environment, there are some basics which need to be explained so that the salient differences between a Conventional war and CI Operations are understood clearly. The environment in J&K and the North East Region is dominated by LIC and its unique characteristics. According to Lt General R K Nanavatty (Retd), ex Army Commander, Northern Command, Internal conflict has been described as under:-

- Internal conflict is extremely complex; its outbreak signals a collapse of the civil government-the administration as well as the law and order machinery. The civilian population may be indifferent and apathetic, if not down right hostile. The adversary is hard to identify and harder to find. Indispensable yet suffocating rules and conditions –often self imposed govern the conduct of operations. The soldier's every action is in the public eye. And the media can be scathing in its criticism.

- It is invariably a long drawn conflict. The people are the objective; the aim is to win their support. The Army can help only create conditions in which the civil government can try to find a peaceful political solution. The Army is not the solution.

- Generally, there is a view that the Army is a reluctant participant in internal security operations which is not its main role; as such operations impinge on the operational freedom of action.

In the conventional war, the battle lines are clearly drawn. The enemy is identifiable. There is no ambiguity about the national goals and the public is not only friendly but supportive. Absence of local population gives full freedom of action. The soldier is regarded as the living symbol of patriotic pride and a saviour of national integrity. Units operate from firm bases, where relaxation in relatively safe environment is readily accessible with fewer lines of communications. Pause between intense phases of fighting, gives time for the soldiers to recoup their energies and they do not come under psychological pressure. While combat stress is inevitable in war, it does not, put serious psychological weight on the minds of the soldiers. Full combat potential is utilised to achieve the military objectives and there are no constraints on the soldiers while fighting the enemy. It is therefore a structured combat scenario.

LIC on the other hand, has become synonymous with a novel and formidable mode of inflicting undeclared war by a weaker nation that does not dare to wage an overt war against her stronger opponent or is substantially intimidated by anticipatory heavy losses involving men and material and/or risk its standing in the world community.

LIC ventures have changed dramatically in perspective and mode of aggression with changing technologies. In modern times, wherein more and more countries are endeavouring to become Nuclear, Biological & Chemical (NBC) warfare capable, war could mean colossal loss of life and resources, with risk to self survival. Hence, LIC has become an ideal and popular choice, to take on adversary till her collapse. It involves mounting terrorist attacks – on the opponent's physical resources of significance, economic power or areas that could inflict heavy casualties on its people making them feel insecure within their own country. It enforces large deployment of troops perennially in roles other than their primary role. It also endeavours to lower the general morale of troops and the civilian population so as to cause unrest and a belief that the government was not capable of providing them with a safe environment.

Today, there has been a paradigm shift in the form of LIC. What was yesterday a low intensity conflict is today a high intensity, no hold bar combat, unleashing gruesome killings, terror, human suffering, anarchy and flagrant violations of human rights. If one were to look at the casualty figures of the civilian population, comparisons defy the classification of low vis-à-vis high. The strategy of "bleeding with thousand cuts" is no less than waging a proxy war. While terrain and living conditions seem a common feature, LIC varies from classical conventional war in many ways. The unique characteristic of LIC is the environment which distinguishes it from conventional war.

Characteristics of LIC. The environment which prevails in LIC has the following characteristics and differs from the conventional wars in many ways as given below:-

- **Presence of Hostile Population.** While in conventional war, the population from border areas is evacuated, in LIC this becomes the centre of gravity as the militants live with the locals and survive like fish in a pond. The Army therefore has to deal with the locals and carryout search operations to weed out the militants from them. These searches impinge upon their personal dignity amounting to their humiliation. The locals may also become victims of the exchange of fire between the militants and the army personnel. The locals know the militants as their own kith and kin and are very sympathetic towards them. The army personnel are viewed as aliens. The Army therefore operates in an environment, where the locals are unfriendly and hostile. The Army could also become a victim of human right violations for inflicting casualties and collateral damage during an encounter with the militants in populated areas. The open aversion of the public to the security forces adversely affects the soldiers. It is indeed humiliating and exasperating to see the very people whom one is out to defend turn against the forces thereby cutting an important feeder of intelligence. It is also disheartening to see the lack of public support and recognition for some stupendous work put in by the army in the affected areas. It is apparent that while the public enjoys all the development work carried out and medical as well as Canteen facilities extended to them, they are hostile from within. It leaves the soldiers with a taste of betrayal.

- **Minimum Force.** LIC by itself indicates a type of warfare / conflict where in the quantum of applied firepower and force is restricted, due to the clause of use of minimum force. The soldier therefore believes that he is fighting with one hand tied behind his back. The soldier is taught to shoot to kill and would do so unhesitant in a conventional war but is restraint to do the same in LIC.

- **Invisible Enemy and Non Recognizable Objective.** In conventional war, there is a clearly identifiable enemy dressed in uniform and a recognizable objective to capture / defend. But in LIC, the militants are dressed like the locals and cannot be distinguished leading to inadvertently killing of an innocent. It is a war between the visible and the invisible. The absence of any physical objective too confuses the soldier because none exists on ground. The absence of a recognizable enemy and objective creates an anomaly in the mind of the soldier.

- **No Boundaries.** Another factor which distinguishes LIC from conventional war is non existence of any boundary. In LIC, there are no boundaries. The threat therefore can emerge from any direction. Apprehension of an attack from any direction puts a sense of fear in the mind of the soldier.

- **Political dominance.** While political objectives would affect military operations, formulation of military strategy and tactics would be the prerogative of the army. In a democratic set up the all encompassing "Political Gains" theory has to be accommodated, thereby numbing the sting of the focus in fighting the LIC. For example, in J&K the Army had to be relocated keeping in view, the wishes of the political leadership whose focus was on the next election.

- **Political Aim.** In conventional war, the political leadership having failed to resolve the problem through political & diplomatic means lets the Army use its might to impose its will on the adversary so as to force a political solution on him. In Conventional war, the political aim is victory, in LIC; the army can at best contain the insurgency.

- **Use of Improvised Explosive Devices (IEDs).** Every soldier is confident of taking on a militant in a one to one encounter and getting the better of him. However, the militants use IEDs which has given a ghost like image to the militants and a capability to strike without coming in direct contact, thus causing frustration to the soldier. Thus the fear of IEDs gives a psychological advantage to the militants.

- **Lack of Mental Respite.** Combating insurgency is a 24 hour job. The level of alertness is very high because the militants can strike anywhere, at any time and from any direction. The routine at the post, Quick Reaction Teams and move at short notice give no respite, time to rest or sleep to the soldiers. This leads to fatigue and exhaustion which have a telling effect on the psyche of the soldiers.

- **Frequent Redeployment.** This is a unique characteristic of CI Operations. The move is not only frequent but to be carried out quickly to reach a particular location resulting in constant redeployment of troops. It is the frequent move at company/

platoon levels that the soldier finds taxing. Large number of troops are sent out as patrols to carry out search operations due to lack of proper intelligence. It gets very frustrating when there is no contact with the militants.

- **Inadequacy of Specialised Equipment.** CI operations require some specialised equipment for detection of IEDs and protection against them. Direction Finding Equipment or Interceptors are required for listening into the communications of the militants. The problem is aggravated because the militants generally use the mobile phones for communication.

- **Fear of Human Rights (HR) Violations.** The fear of perpetuating Human Rights violations and subsequent judicial harassment inhibits the actions of the men and also make them vulnerable to the militants during operations. The worst is when the higher commanders fail to hold the hands of the subordinates during a HR violation. The problem gets further aggravated when the media blows the incident.

- **Lack of Recognition.** There is a general feeling that unlike the conventional war, due recognition is not being given for the services rendered by the soldiers in CI Operations. While the soldier feels that he is giving every thing for maintaining the integrity of the nation, no body has a word of praise for him when he steps out into the streets.

- **Frustration Due To No Results(Numbers' Game).** There are times when there is no contact with the militants. When this goes on for days and months, the frustration sets in. It does not matter how much one slogs, what matters is the number of kills and weapons recovered.

- **No Mistake Syndrome.** When operations are on, reverses is a natural fall out. But this is not acceptable and formal inquiries are ordered. This has a negative impact and makes commanders not only cautious but also play safe. This kills the spirit of aggressiveness in the soldiers.

- **Political Solution to the Problem.** While in conventional war, the political leadership having failed to resolve the problem through political and diplomatic means, lets the army use its might to impose its will on the adversary so as to force a political solution on him. In LIC, then is no military solution to the problem. The army can, at best, contain the insurgency. Its resolution can only be achieved through political means.

Nature of LIC Operations. Major General Arjun Ray in his book 'Kashmir Diary' has given the right description of the nature of LIC and counter insurgency operations.[19]

- Counter insurgency operations whether-winning or losing are an uphill and heartbreaking task.

- The battlefield appears to be frozen in time. While one is running all the time, it appears that one is standing at the same place.

- The levels of uncertainty and friction, at times, become intolerable. A time comes when one is tempted to give up.

- A stalemate in the situation leads to a battle of wills; the will of the security forces to maintain pressure and the will of the militants to sustain it.

- The operations are continuous, there are no tactical or logistics pauses and no time outs or half times.

---

[19] Kashmir Dairy by Maj Gen Arjun Ray

- The frustration mounts when there is no break through but the expectations to perform and produce results are high (Numbers Game).

- It is warfare where execution is decentralised. The impact and influence of higher commanders is less as their orders and intent take time to reach the subordinate commanders particularly at the execution level.

- The soldiers are caught in a cross fire, moral vis-à-vis immoral and the dividing lines are blurred.

Other dimensions of LIC are :-[20]

- Time seems to favour the militants. The longer the struggle, the more it goes in their favour because they get bloodied and start having a better comprehension of the tactics, modus operandi and the vulnerabilities of the security forces.

- The local people get more and more alienated because of the harassment suffered due to ongoing operations of the security forces. Casualties to the militants too become a cause of alienation as any militant killed, wounded or apprehended is related to some body or the other. The international community too starts sympathising with the militants and supporting their cause.

- No timeframe can be fixed for resolving such complex issues. Any military leader who professes that he has got across the hump and sets a date for wiping out insurgency is not only indulging in self deception but also sending wrong signals to those in authority for taking certain vital decisions and formulating a policy on the subject.

---

[20] Tackling Insurgency and Terrorism by Maj Gen Samay Ram(Retd)

- There is no clear mandate from the political leadership. In the absence of such a mandate, the army hierarchy formulates its own strategy and undertakes the onus of resolving a complex political problem militarily.

- The army ends up containing the situation putting the political leadership to complacency in resolving the problem through political and diplomatic means.

- LIC is a very complex issue and there are no easy solutions to it. The problem is further complicated by the mushrooming of a large number of militants with different ideologies and goals making it difficult for the government to choose the right group for negotiations. Since some of the major groups are controlled by the hostile neighbour, mutual agreements make no headway. The conflict thus remains a protracted and long drawn affair with its attendant human and financial cost. From the military point of view such a protracted conflict, places a psychological weight on the military mind.

Environment in LIC

The deployment for CI Operations is generally based on a grid system entailing large areas to be dominated. A unit of RR has nearly 6-8 company size posts. The operational commitments are heavy involving 10xBattalion level and 10xcompany level operations by night in a month. The operations are conducted mostly at sub unit level thus requiring large number of officers at the junior level – in the absence of adequate number, the burden is shared by those few posted. There are frequent moves and redeployments. The living is therefore under ad hoc conditions. There is a great need for constant vigil as militants can attack any time and from any direction. Lack of actionable intelligence puts great pressure on the troops to seek contact with the militants. The conditions of terrain and climate are harsh with hardly any resources available. Troops get inadequate time for rest and sleep; with no mental respite instilling a feeling of isolation and fatigue. Being separated from the family, the men are constantly concerned about their well being, especially in the event of his being killed in action. Instinct for survival is high. The fear of an IED blast or a suicide attack is constantly on the mind. The pressure to perform is high but the search operations are frustrating. Everyone is afraid of co lateral damage during operations. There are no visible signs of victory and an end of the problem in view. The soldier remains worried about coming back and battling the same situation.

Environment in Peace

The unit returns to a peace station with a lot of hope for rest and rejuvenation. The men find themselves in a good environment - better living conditions and good quality of food. The life is more relaxed and nights are peaceful. Families join and there is hope of being together. But soon the charm wears off and the men find themselves in another type of atmosphere, where there is no danger to life but no respite either. The routine is laced with a series of events - sports competitions, training competitions, individual training cycle, field firing, collective training and if nothing else social events like welfare meets, Barakhana and so on. Above all, there are a series of inspections and VIP Visits. Thus the man finds himself caught up in series of activities that give him little quality time to be with his family. Soon the charm of a peace station wears off and the men start feeling that they were better off in CI operations. Soon the time passes by and the unit is once again ready for field tenure, high altitude area or in the North East again battling the militants. This cycle of induction, de induction, and re-induction goes on. Soon the man realises that his time is nearing to go home and start another career, even though he has yet to settle his children before retiring.

Conclusion

A perusal of the environment; whether Field or Peace is quite taxing to the soldier. The environment especially in CI Operations is very demanding. It is considered as a war of nerves and puts a great psychological weight on the minds of the soldiers. Stress, therefore is an inevitable fallout.

CHAPTER 4 -STRESS AND CAUSATIVE FACTORS (STRESSORS)

Occupational Stress. "It is the result of interaction of work conditions with characteristics of the worker such that demands of the work exceed the ability of the worker." -Rose and Altair - (1944)

The peculiar nature of LIC and its intensity of violence over a protracted period, by itself, are the major factors of stress. However, various studies on the causative factors of stress have shown that there are several other factors which contribute to stress. Since stress is caused by Stressors, it would be prudent to identify the same through various studies on the subject as most of these studies are based on scientific approach to the subject.

Focus of Study

The focus of these studies has been the NE region and J&K. Some of the studies have brought out that both these regions have peculiar environments- which have their own characteristics to distinguish one from the other. It is therefore vital to understand this distinction.

Extracts of Study Reports

The following studies/Reports have been evaluated

- DIPR Study, Refer to **Appendix C** attached.[21]

- DRDO Study, Refer to **Appendix D** attached.[22]

---

[21] Extracts of DIPR Report by Ravi Sharma. *The HINDU dated 08 Aug 2007*

[22] DRDO conducts study on Stress by Ravi Sharma- *he HINDU dated 08 Aug 2007*

- Psychological effects of LIC Operations by Suprakash Chaudhary, D S Goel and Harcharan Singh. Refer to **Appendix E** attached [23]

- Standing Committee on Defence (2008-2009) [24]

- Evolving Medical Strategies for Low Intensity Conflict by Lt Col Ajay Dheer, et al.

- Stress Management in CI Operations by Maj R Rajan-THE INFANTRY, 2004.

- In addition, data discerned during interaction with units/ formations in J&K and feed back from ex formation commanders.

Examination of Studies/Reports

Now let us consider what the above reports have to say.

DRDO Report. There are three main Operational Stressors. These are: Fear of Torture, Uncertain Environment and Domestic Pressure.

Study on Evolving Medical Strategies for LIC By Lt Col Ajay Dheer et al. Combat Stress is caused because LIC is a campaign of nerves, less Military and more Psychological. The mental strain of fighting one's own people is infinitely greater than fighting the enemy, with hands tied behind the back constraints. Environmental Stressors which adversely affect a soldier operating efficiency in LIC are – "Poor living conditions, adverse weather conditions, and extended tenures beyond 18 months". (*This means that the two main causes of Stress are Operational and Environmental Stressors*).

---

[23] Indian Journal of Psychiatry, 2006.

[24] 31st Report on Stress Management in the Armed Forces, Lok Sabha Secretariat, Oct, 2008.

DIPR Report. The No1 Cause for all Ranks- Occupational Stressors, No 2 Cause for Officers and JCO's - Familial Matters and for OR's - Lack of Basic Amenities. No3 Cause – For Offrs- Miscellaneous Factors like lack of Local support, Zero error Syndrome and lack of sophisticated weapons/equipment, For JCO's it is lack of basic Amenities and OR's- Familial Issues. General Causative Factors like Increased Work Load, Non grant of Timely Leave, Lack of Adequate sleep, Lack of Basic Amenities, Poor Pay and Allowances and Pressure from the Family Front. The report also attributes Poor Mental Robustness and Alcoholic Dependency as the next two factors.

Standing Committee on Defence.

- **Socio- Economic causes(** Family Discords/Disputes-Inability to take care of separated family-Extra Martial Affairs-Poor financial condition, Debt/Dowry demand-Sexual exploitation, Alcoholism and Drug abuse)

- **Environmental /organizational(** Prolonged separation from family-Prolonged commitment in LIC-Uncongenial and stress environment- Frugal living conditions and lack of rest –Low relationship and high task orientation leaders and denial of leave in an emergency)

- **Medical/ Psychological.**(Mental Depression-Poor interpersonal relations-stigma attached for Psychiatrist help-Incurable disease and Sexual dysfunction)

   (Surprisingly there was no cause attributed to Operational/ Occupational Factors)

Psychological affects of LIC by Lt Col S Chaudhary et al.

- **Occupational/ Operational Factors.** (Anger at fighting with constraints-Anger at public admonishment-bitterness at dealing with the Jamayatis)

- Feeling of uncertainty-Futility of CI operations-Fear of ever present danger.

- **Social Factors.** (Lack of co-operation/ hostility from local population- Adverse publicity in media-Difficulties in rail travel-Hostile attitude of human rights groups-Feeling of insecurity regarding families).

In the article on Stress Management in CI Operations, Maj Rajan has brought out that there are Physiological, Psychological and Organisational factors as the main causes of Stress.

Interaction with Units/ Formations.

- No I Cause   -   Physical/ Mental Stress* *(Environmental Stressors)*

- No 2 Cause   -   Family Pressures

- No 3 Cause   -   Kill Maximum Terrorists

- No 4 Cause   -   Poor Living Conditions*

- No 5 Cause   -   Own Safety.

According to Lt Gen Patankar, ex Corps Commander, 15 Corps, the likely causes of stress are:-

- Domestic Worries.

- Marital Discord.

- Financial Liabilities.

- Superssession.

- Living up with the Joneses.

- Long working hours with insufficient time for Rest and Sleep.

- Degree of difficulty of tasks allotted(Particularly if skill levels are inadequate to meet the challenge)

- Fear of Failure/ Loss of face.

- Lack of Recognition or Reward.

   (The above mentioned causes fall under the Domestic, Environmental or Operational Stressors.)

The author had the opportunity to interact with 50 cases of Mental Disorder in a hospital. Some interesting observations made by the author are given as under:-

- **Patient A.** He was Naik, acting as the Kote NCO of his company. As such the Treasury chest was also under his custody. There was a sum of Rs 18,000.00 in the Treasury Chest. One day he came to the Kote and opened the same for drawl of some items. Having done so, he locked the kote and went to hand over the items. After some time the Company Commander called him and asked him to bring some cash from the Treasury Chest. When he went there and opened it, he found the lock open and Rs 6,000.00 missing. He immediately reported the matter to the Company Commander. The Company Commander accused him of stealing/ mis- appropriating the same. The Naik denied having taken the money. A Court of Inquiry was held in which he was blamed for stealing the amount. The matter was brought to the notice of his Commanding Officer who asked him to return the amount. He again denied having taken it. No body listened to his pleas. He was mentally very disturbed for being called a thief in the unit. The unit commanding Officer seeing his behaviour filled up AFMS-10. He was evacuated and sent to the hospital for investigations. The Psychiatrist found him normal.

- **Patient B.** This jawan was from a RR Battalion. On completion of his two years tenure, he requested to be sent back to his parent unit. But the unit did not do so as his relief had not arrived. He requested for an interview with his Commanding Officer. He too refused to revert him to his parent unit. He became obsessed with the desire to go back and became mentally disturbed. The Commanding Officer filled up the AFSM-10 Form and sent him for investigation. The Psychiatrist found nothing wrong with him and declared him as fit.

- **Patient C.** This was a case of being jilted in love. The girl he loved was married off by her parents to some one else. He could not bear to part with her and became a case of mental disorder.(There were three such cases of similar nature).

- **Patient D.** This was a case of supersession. A Havildar was not promoted to the rank of a Naib Subedar as he could not pass the promotion test. This Havildar could not bear it because he was worried about what the members of his family and his colleagues would say. He became a psychologically disturbed person.

- **Patient E.** This individual wanted to be an officer but had failed to make it in the Services Selection Board. He thought he was better qualified than the officers. He conversed in English to show that he had the requisite qualifications. His obsession to become an officer made him a case of mental disorder.

- **Patient F.** This was a case of heavy debt which he could not return. He had spent a lot of money on the treatment of his father who had lung failure. His inability to pay the heavy debt stressed him so much that he became a psychologically disturbed person.

- **Patient G.** He had got married only two years back. He had a very nagging wife who was constantly quarrelling with him.

He could not leave her because of the pressure from parents and relatives as well as the social stigma attached to it.

- **Alcoholic Cases.** There were three cases of alcoholism All of them appeared quite sturdy and healthy. They did not disclose how they managed to get extra drinks when the consumption in the units was controlled through the sale by unit CSD Canteen.

From the analysis of the Psycho/Alcoholic cases, it could be discerned that the causes of Mental disorder and alcoholism were due to:-

- Domestic worries.

- Marital Discords.

- Jilted Love.

- Financial Liabilities/Debt.

- Suppersession.

- Loss of Face.

- Misplaced ambition not compatible with the capabilities.

- Not relieving in time on completion of tenure.

- Long working hours with insufficient time for rest and sleep.

The important observation was with regard to the use of AFMS-10. Since some cases were found NAD (Not Detected), was this form being misused by the officers to get rid off cases which they could not handle or was this an easy route for some individuals to escape the rigours of life in the unit?.

Evaluation of Studies/Reports

From the analysis of the data, it is evident, that there are different stressors as enumerated below:-

- Operational/ Occupational stressors.

- Environmental/ Organisational stressors.

- Domestic/ Familial stressors.

- Inter personal stressors.

- Psychological Stressors including Medical Factors.

It is not necessary as to what grouping a particular stressor falls under. The grouping is more for convenience, so that the focus is on each factor separately. Secondly, it has been observed that every individual has his degree of threshold depending upon his personal background or the type of training he has been put through. Soldiers from a poor background and having undergone tough training seem to have a higher threshold.

Analyzed data of the above studies shows that the following factors cause stress:-

- *Operational/Occupational Factors*. Anger at fighting with constraints-Zero error syndrome-Less sophisticated weapons- Feeling of uncertainty-Fear of ever present danger/ unexpected attack and kill maximum militants.

- *Environmental/ Organisational Factors.* Adverse publicity in media- Lack of support from local population/Anger at public admonishment-Hostile attitude of Human Rights groups-Lack of basic amenities including poor uniform— Uncongenial/ stressful environment like poor living conditions- Pressure of work and inadequate time for sleep and rest.

- **Domestic/ Familial Factors.** Financial and Personal issues-Family matters including marital discord- Family diputes/ discords-Financial pressures like debt, dowry and inability to support separated family-

- **Medical/ Psychological Factors.** Incurable disease like Sexual Dysfunction, Sexual exploitation and stigma attached for seeking psychiatrist help and physical help.

Considering various studies/the results of Interaction with the troops and Analysis of the Stressors above, it is concluded that in LIC environment, the two main Stressors (No1 and 2) are Occupational/Operational and Environmental/Organisational Stressors. They are both interdependent and the most important causative factors of stress. Calling a stressor as No.1 or No.2 is purely a matter of semantics. The No 3 Stressor is Familial Pressures. No 4 is the Medical/ Psychological factors. What matters are the issues that cause Stress and not the grouping under which they fall.

The above is in consonance with what Paramahansa Yogananda has to say about Stress,

"The disturbance of mental equilibrium , which results in nervous disorders , is caused by continuous status of excitement or excessive stimulation of the senses, indulgence in constant thoughts of fear, anger, melancholy, remorse envy, sorrow or worry and lack of necessities for normal happy living such as right food, proper exercise, fresh air, sunshine, agreeable work and purpose in life, all are the causes of nervous disease".

Stress is also the result of inherent contradictions due to individual expectations and organisation culture, working environment and individual factors. This argument has been supported by a quote as under :-

"These days it seems all of us are too serious. Understand the link between your expectations and frustration level. When ever you expect something to happen your way and it isn't, you are upset and you suffer. But when you let go of your expectations, accept life as it is, you are free. To hold on is to be serious and uptight. To let go is to lighten up." Richard Carlson

In Peace locations, since there are no operations and the environment is comparatively more conducive, Domestic Stressors would become the No. 1 Stressor followed by the Environmental (Pressure of Work) as the No. 2 Stressor.

There is no consensus of opinion with regard to the stressors.

Stress levels are different for Different Ranks, Different Areas and Different Ages.

Stress Levels.[25] Stress levels in LIC and other areas is shown in tabulated form below:-

Table 2- Stress Levels

Ser No	Stress levels	LIC Areas	Other Areas	Remarks
1.	Severe	0.53%	0.02%	Significant
2.	Moderate	8,45%	3.87%	
3.	Mild	66.3%	53.5%	NO STUDY ON PTSD
4.	Sub clinical	24.6%	42.76%	

---

[25] Table 3- Psychological effects of LIC operations, Lt Col S Chaudhary et al

Anatomy of Stress

In the Environment the soldiers operate and the job they have to do, they face varied challenges. For example:-

- In LIC Environment, the issues are- Occupational/Operational Factors – (Pressure to perform that is, Killing Maximum Militants– Fighting under constraints- Zero Error Syndrome- Unsympathetic Local Population and Adverse Media.) Environmental Factors (Inadequate time for Rest/Sleep-Poor quality of food and Habitat- Paltry Pay and Allowances), Domestic/ Personal Factors- (Separation from the family- Family conflicts- Heavy emotions that accompany a broken heart or death of a loved one).

- In Peace stations, besides the Domestic/Personal Issues, there is a pressure to perform stressful jobs-pressure of too many Training, Social and Sports events, leaving hardly any time to spend with the family.

- These challenges trigger the coping mechanism in the soldiers. And they try to cope with these challenges using any of the three strategies- act- change or withdraw with the aim to Change, Maintain or Adapt.

- This coping process requires investment of energy. It is continuous and ongoing.

- When there is a failure to cope with such challenges, that is, when the demands of the External Environment exceed a soldier's resources for coping, it would result in stress.

- Since the degree of coping is different for different individuals, one finds different stress levels for different Ranks and Age Groups.

This is how it affects:-

- In the Initial stage, the soldier has a europhic feeling of encounter with the new Assignment /Challenges.(Honey moon stage)- Sub clinical Stage

- Energy resources gradually deplete. Habits and strategies for coping with stress formed. (Dysfunctional Features)- Mild/ Low Stage.

- There is an over drawl of resources. Realisation comes too late that the energy resources are limited. (Fuel Shortage stage). - High Optimum/Mild Stage.

- The physiological system becomes so devastating that it can completely knock down a person. With all adaptable energy depleted, one may lose control over one's job (Hitting the wall Stage) - Severe Stress.

- The consequence of high level job stress leads to personal frustration and inadequate coping skills which has many Personal, Organisational and social cost- Pain (Burn out Stage) – Acute Stress.

The most typical Syndrome of the Dysfunctional is withdrawal from the situation. The soldier ceases to react to his surroundings and in fact shows no ability to perform any coping or adaptative behaviour. In such a situation, the individual adopts a "couldn't care less attitude". The onset of this syndrome rarely follows the first exposure to battle, but develops after continuous exposure to combat and with a frequency that increases relative to the intensity to the duration and intensity to the exposure. Several days are required before the threshold of this symptom is reached. The total amount of exposure to stress determines the point of failure to cope. The total amount of stress determines the point of failure to cope. Intense stress over a short period has a similar effect to milder stress over a more extended period.

The most common failure to cope in the field of battle is the Neuro-Psychiatric Dysfunction often described as "Battle Fatigue"- Shell Shock or Combat Stress.

Conclusion

The war of LIC is a war of nerves. It has more psychological dimension than military. It causes Stress on all ranks through various Stressors, be they, Occupational, Environmental or Family Pressures. The Army must look into these factors and take remedial measures to address them.

Appendix C

(Refers to Chapter 4, Page 57)

EXTRACTS OF STUDY- DIPR

1.　　A designated team of scientists from DIPR carried out study on the problem of Stress, Suicides and Fratricides in the Army. The team visited 3 and 4 Corps in the North East Region and 15 and 16 Corps in J&K. They interacted with Officers, JCOs and ORs and interviewed a total of 2001 Personnel.

2.　　According to the report, Occupational Factors are the No1 Cause for Stress. While the No. 2 cause of Stress for Offrs and JCOs was Familial matters, for the ORs it was the lack of basic amenities. The No. 3 cause was as under:

- Offrs　　　-　　Miscellaneous factors like Zero Error syndrome and less sophisticated weapons.

- JCOs　　　-　　Lack of basic amenities.

- ORs　　　-　　Familial issues.

3.　　The next two reasons were Poor Mental Robustness and Alcohol dependency.

4.　　The general causative factors were; increased work load, non grant of timely leave, Poor Pay and Allowances and Pressure from family front.

5.　　Command wise, the findings were as under:-

NORTHERN COMMAND

6.   **Officers.**

- No 1 Cause   -   Occupational Factors.

- No 2 Cause   -   Staying away from family.

- No 3 Cause   -   Lack of basic amenities.

7.   **JCOs.**   Lack of basic amenities - familial issues.

8.   **ORs.**   Low Pay and Allowances, lack of basic amenities like woollen clothing and uniform and Harassment by senior officers.

EASTERN COMMAND

9.   No 1 cause- Occupational Factors.

10.   Other Causes- Lack of basic amenities, Poor quality of uniform, poor living conditions and no transport for going back.

Rank wise

11.   **Offrs.**   Personal causes as prominent precursors.

12.   **JCO's & OR's**. Occupational - Familial issues more important.

Both Commands

13.   Perceived Humiliation by higher ups as the significant trigger.

14.   Officers listed lack of cordial relations between leader and led but JCOs and ORs thought these to be inadequate basic amenities and occupational factors.

Appendix D

(Refers to Chapter 4, Page 57)

EXTRACTS OF DRDO STUDY REPORT

DRDO Conducts Studies on Stresses on Indian Soldiers

(Dated 21/11/2007)

Defence Institute of Psychological Research (DIPR), a laboratory of Defence Research and Development Organisation (DRDO), had conducted two studies. One of the "Psycho-Social aspect of optimizing the operational efficiency of Security Forces to combat insurgency" during September 2000 May 2005 in the North Eastern Region. The major outcomes of the study are:-

•   Three main operational stressors, like fear of torture, uncertain environment and domestic stresses are responsible for most of psychological problems in various groups of Armed Forces. Middle rank officers as compared to Jawans and Junior Commissioned Officers (JCO) were found to be more vulnerable and stressed out.

•   Mental disorders in the form of Post Traumatic Stress Disorder (PTSD) have been observed in traumatized troops, which form the basis for various somatic symptoms. About six to nine months after the detention of captured militants is the most appropriate period to bring about a change in the attitudes of youth and influence their minds towards national identification.

•   A strong need to inoculate and orient officers by undertaking Combat Stress Management training programmers. Need for immediate therapeutic intervention during post-traumatic stress disorders jointly by unit leaders and a professional psychologist.

There is a strong need to carry out an appropriate need analysis with respect to demographic variables of a particular region.

•   Another study was carried out in December 2006 on Suicides and Fratricides among troops deployed in counter insurgency areas and suggested a number of remedial measures to deal with stress related issues in the Army. Some of the suggestions are: liberalized leave policy, deployment of psychological counselors in counter insurgency areas, periodical review, monitoring and analysis of stress related incidents, organisations of stress management training programmers and disseminations of reading materials on stress management in appropriate languages. The remedial measures to manage stress have been implemented in the entire Army, both in field and peace areas. This information was given by the Defence Minister Shri AK Antony in a written reply to Shri Motilal Vora and Shri Satyavrat Chaturvedi in Rajya Sabha today.

(*THE HINDU - 08 AUG 07*. REPORT BY RAVI SHARMA)

Appendix E

(Refers to Chapter4 Page 58)

EXTRACTS OF REPORT BY LT COL S CHAUDARY et al

Table 1.   Negative and Positive Determination of Morale (Stressors)

S.-No	Negative Determinants	Subjects in LTC (n=568) (%)	Subjects in Other Areas (n=668) (%)	Chi square value	P
Operational Factors					
1	Anger at fighting with constraints	79.75	59.68	54.19	<0.01 S
2	Anger at public admonishment	77.11	68.13	11.52	<0.01 S
3	Bitterness at inabliity to deal with 'jamayatis'	59.33	36.62	15.73	<0.01 S
4	Ambiguity regarding aim	26.94	27.11	0.00	>0.95 NS
5	Feeling of uncertainty	24.82	17.43	9.32	<0.01 S
6	Futility of counter-insurgency operations	23.77	15.67	11.70	<0.01S
7	Fear of ever present danger/ unexpected attack	20.07	13.38	9.13	<0.01S

Table 1. Negative and Positive Determination of Morale (Stressors)

S.-No	Negative Determinants	Subjects in LTC (n=568) (%)	Subjects in Other Areas (n=668) (%)	Chi square value	P
Social factors					
1	Financial dissatisfaction	54.23	58.45	2.06	>0.10 NS
2	Lack of cooperation / hostility from local population	53.17	38.56	24.42	<0.01 S
3	Lack of societal support	50.53	46.66	1.55	>0.20 NS
4	Disgust at a corrupt polity	47.71	46.12	0.28	>0.50 NS
5	Adverse publicity in media	43.49	31.87	16.33	<0.01 S
S-Significant , NS-Not significant					

CHAPTER 5- SUICIDES IN THE ARMY

The issue of suicides and fratricides in the army has become a matter of serious concern. The union Defence minister, Shri A K Antony, had to order a study by the director, Institute of psychological research, to ascertain the causes of this malady and suggest measures to arrest this trend. The Army, itself, has been seized with the problem and so also other well-meaning people who find it difficult to reconcile between their perception of the Indian soldier and the harsh realities.

This chapter deals with the causes of "Suicides" in the army.

Suicide [26] (Latin- Sui caedre-to kill one self) is a life threatening act with a conscious intent to end one's life. It is diverse in pathology and has varied manifestations. It is an old age phenomenon and all religions view it as a sin. In the Army it is considered as an act of cowardice. Besides being a social stigma, it results in loss of precious life of a trained soldier. Hence, it is imperative to ascertain its causative factors. Unfortunately accurate predictions of this act are not possible. The best that can be done is to identify high risk groups and keep a watch on them. Such individuals are psychiatrically ill with problems like depression, alcohol dependence and schizophrenia, those with personality problems and those who have made suicidal attempts in the past. Suicide attempts are the outcome of Environmental stressors in the Domestic/Societal and occupational

---

[26] Extracts on Suicides from Wikipedia, the Free Encyclopedia.

spheres. These stressors result in the failure of the normal coping mechanism and the adaptative ability of the individual, leading to taking of the extreme step of ending one's life. If such stressors can be managed then the suicidal problems could be addressed.

Suicide cases are on the increase globally and are a major cause for loss of lives. Therefore, prevention of suicidal acts in the Army is the need of the hour especially as the vulnerable age group of the country constitutes the majority of the Armed Forces personnel. Modern medicines view suicide as a mental health concern, associated with psychological factors such as coping with depression, inescapable fear or suffering (being sexually abused); or other mental disorders and pressures. Suicide is interpreted in this framework as a cry for help and attention or to express despair and wish to escape rather than a genuine intent to die. *Suicide is also an expression of internal aggression.*

Understanding the Phenomenon of Suicide

Secret of Suicide. While carrying out the autopsy of suicide cases, an International Research group led by Micheal Poulter at the university of Western Ontario and Hymie Anisman of Carleton University found elevated levels of a particular protein in the part of the brain that controls anxiety and stress.They found that this protein caused a chemical modification resulting in the shutting down or malfunctioning of the anxiety and stress controlling gene impairing the individual's ability to handle Anxiety and Stress. The nature of this chemical modification is long term and hard to reverse. This fits in with the cause of depression that prompts a person to commit suicide.

How does a person reach the stage of putting an end to his life? This issue is analyzed as under:-

- There are some individuals with sulcidal tendencies or Psychiatric disorders (Anxiety and Depression). When faced with a problem, while some individuals try to resolve it, others attempt an escape route and take to Alcohol- even becoming Alcoholic..

- Depending upon the Environment, some of them could already be under stress. The gestation period therefore depends upon the environment and the gravity of the problem.

- In such a process, if an individual realizes that the problem cannot be resolved, or there is no recovery from the disability and the ailment; such an individual gets mentally so disturbed that he goes into physical and emotional agony leading to depression.

- When such a realisation comes, there is a shutting of the Anxiety and Stress controlling gene as explained above. This impairs the individual's ability to handle anxiety /stress and he or she commits suicide.

Fact Data Sheet

The Fact data sheet with regards to suicides is given below. The data is sourced from various sources and also that which was obtained during interviews with the veterans/interaction with units and formations. The concern is genuine as the army is losing approximately 100-120 men per 100,000 in suicides. This figure per lac is otherwise much less than those of National average and of other armies of the world which are given below:-

(a) National   -   11

(b) US Army in Iraq   -   17

(c) UK Armed Forces   -   14   (Based on data
                                        2004 to 2006)

(d) French Armed Forces   -   18.2

(e)   Sri Lankan Army   -   15.1

Some people take solace in the fact that the cases of suicides are much less compared to those of the US and British army which are operating in Iraq and Afghanistan. Such people have to understand that the background, the psyche and the resilience of the US/ British soldiers is quite different to those of the Indian soldier. The comparisons are wrong because there is no match to the Indian soldier's sense of discipline and capacity to brave hardship. Therefore, given the fact that the Indian army is a well trained and disciplined force, the figures of suicides and fratricides are a matter of concern. There is therefore, a need to address the causes and find remedial measures so that we save avoidable casualties.

As per WHO, there are one million suicides per year (one death every 40 Secs-one attempt every 3 Secs) [27]

SUICIDES
A MOUNTING TOLL - NATIONAL LEVEL

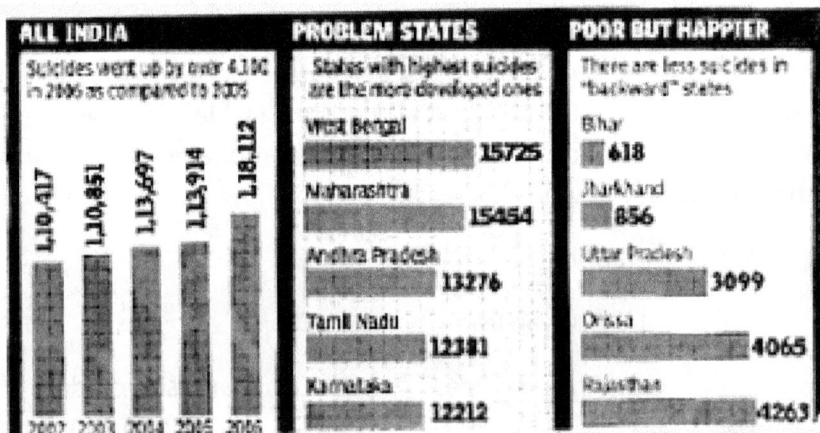

ALL INDIA	PROBLEM STATES	POOR BUT HAPPIER
Suicides went up by over 4,100 in 2006 as compared to 2005	States with highest suicides are the more developed ones	There are less suicides in "backward" states

ALL INDIA: 1,10,417 (2002), 1,10,851 (2003), 1,13,697 (2004), 1,13,914 (2005), 1,18,112 (2006)

PROBLEM STATES:
West Bengal — 15725
Maharashtra — 15454
Andhra Pradesh — 13276
Tamil Nadu — 12381
Karnataka — 12212

POOR BUT HAPPIER:
Bihar — 618
Jharkhand — 856
Uttar Pradesh — 3099
Orissa — 4065
Rajasthan — 4263

---

[27] A *Times of India* Report.

In India, suicides went up by 4100 in 2006 as opposed to 2005 reflecting a rising trend. West Bengal, Maharashtra and Andhra Pradesh have maximum suicides.

- There are fewer suicides in backward states like Bihar, Jharkhand and even Uttar Pradesh (UP) (BIMARU States). Even UP with 16.5% of India's population has only 3% of the National toll.

- Total Suicides per day=312 (Men 64% and Women 36%)

- Causes- Family Problem (22%) - Illness (22%)-Failed Love Affairs and Debt (3%) each- others (50%)

As per the data published in the CLAWS research paper on "Challenges of Management and Combat Stress" (forthcoming), suicide rate in India is 11 per lac, 2nd in South East Asia and 45th globally. Kerala has the highest rate of 29 per lac and Bangalore contributes 70% of total suicides in the country.

(The figures will keep changing every time the survey is carried out. The basic issue remains that rates of suicides in India are very high. The states having higher rates of literacy have more suicides than those with low literacy rates)

Total suicides in the Army from 2003 to 2008 are as under-*

Suicides In The Army

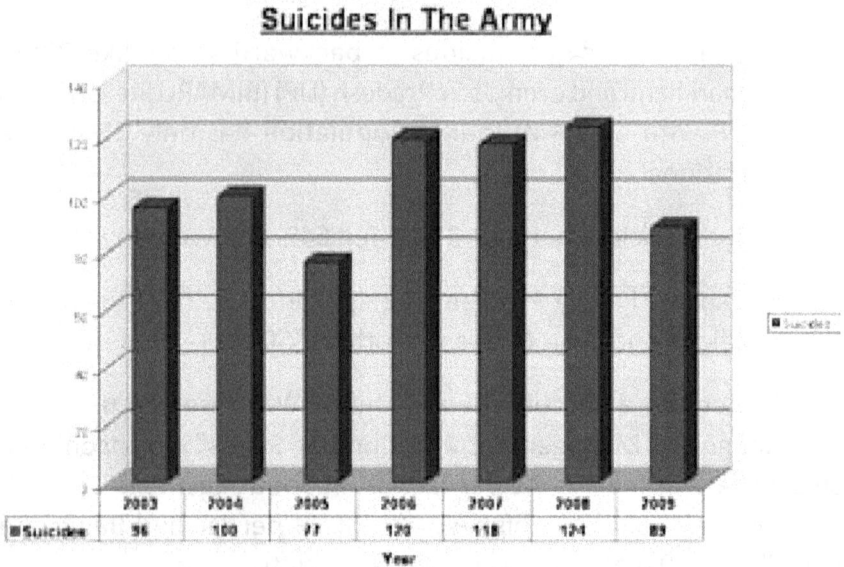

| Suicides | 96 | 100 | 77 | 120 | 118 | 124 | 89 |

- **Average=106 per year**

Army- National. Army suicides per lac were more than National Average in 2004 but less than the national Average in 2007.

Army- Other Services. Average per lac is less than IAF (15) and BSF (11) but more than the Navy (6)

Peace - Field. (Data based on suicide cases in 2006-120). Army has more number of suicides in Peace (75) than Field.(45)- 63%:37% But on pro-rata basis there are more suicides in Field (13 per 10,000) than in Peace (11 per 10,000).

Place	Officers	JCO's	ORs	Total
Field Area	3	1	41	45
Peace	1	3	71	75
Total	4	4	112	120

- **ERE- Parent Unit.** ERE-14.1%- Parent Unit- 87%.

- **Duty- Leave Station.** Duty -96%- Leave Station- 4%.

- **Age wise**. Maximum suicides in 20-25 Age Group. Minimum in above 30 Years age.

- **Service Group.** Maximum Risk Group is between 7-10 Years service group.

- **Marital Status-(Data based on suicide cases in 2006)**

  Married Vis-à-vis Unmarried (%) = Offrs-50:50, JCO's-100:0, ORs-55:45

Category	Offrs	JCOs	ORs	Total
Married	2	4	66	72
Unmarried	2	0	46	48
Total	4	4	112	120

Urban- Rural. 1: 3.

- **Region wise.** Northern- Southern-Western – Eastern and Central.

- **Arms/ Services Wise**. Based on the data of 120 suicides in 2006, On Pro -rata basis, the maximum suicides were in the RVC- AOC and DSC. As per the findings in CLAWS paper on "Challenges of Management and Combat Stress Management in LIC Environment"(forthcoming), highest number of suicides are in the Artillery followed by AMC, Inf, Engrs and Sigs.

Modus Operandi. There are several ways of committing suicide but the commonly used modus operandi used in the Army is as per details given below:-

Modus Operandi	%
Use of weapon	35
Hanging	53
Poisoning	7
Jumping	2
Miscellaneous	3

Causes of Suicides

Based on the various studies/reports, their evaluation and interaction with the veterans as well as units/ formations, it is very difficult to ascertain the exact cause(s) of suicides particularly so when the evidence is either not forthcoming or is destroyed deliberately. Not withstanding the problems being faced, it has been possible to identify the following major causes of suicides:-

- Domestic/Family Problems   -   64%
- Organisational Factors   -   18%
- Medical/ HIV   -   16%
- Self Respect   -   2% (Victim or perpetrator of sexual Abuse)

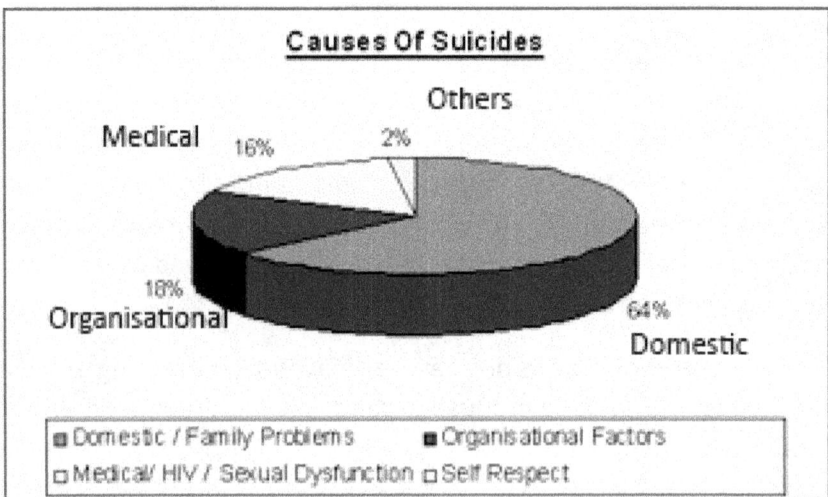

Causes Of Suicides

Another view on the subject is reflected graphically in the CLAWS paper on "Challenges of Management and Combat Stress in LIC Environment . The conclusions are generally the same.

CHART DEPICTING REASONS FOR SUICIDE[28]

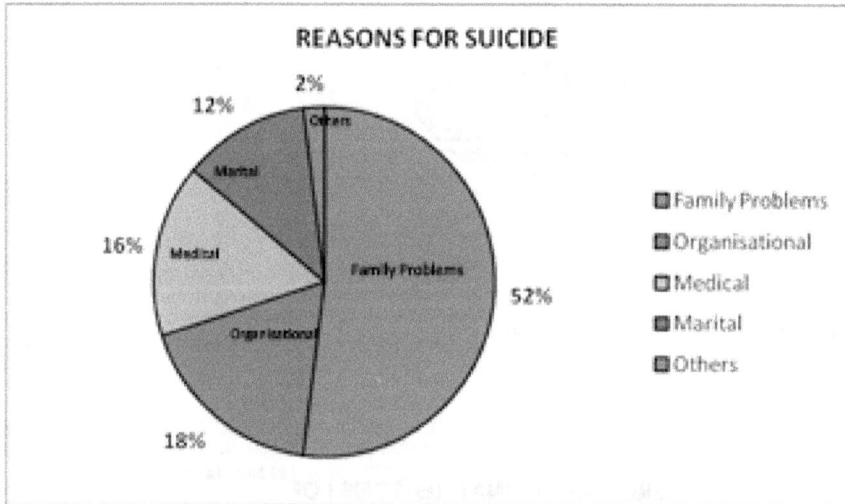

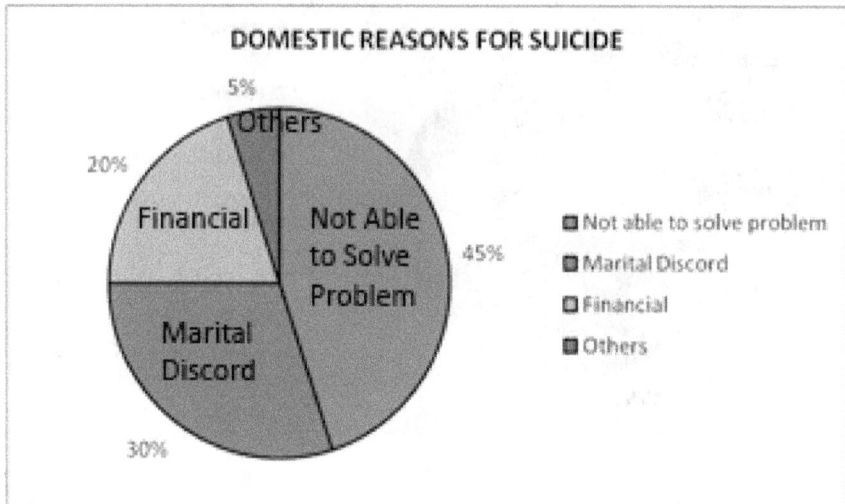

REASONS FOR SUICIDE

2%
12%
Others
Marital
16% Medical
Family Problems
52%
Organisational
18%

- Family Problems
- Organisational
- Medical
- Marital
- Others

DOMESTIC REASONS FOR SUICIDE

5%
Others
20%
Financial
Not Able to Solve Problem
45%
Marital Discord
30%

- Not able to solve problem
- Marital Discord
- Financial
- Others

---

[28] Challenges of Management and Combat Stress-CLAWS Paper.

REASONS FOR DISSATISFACTION IN FAMILY LIFE

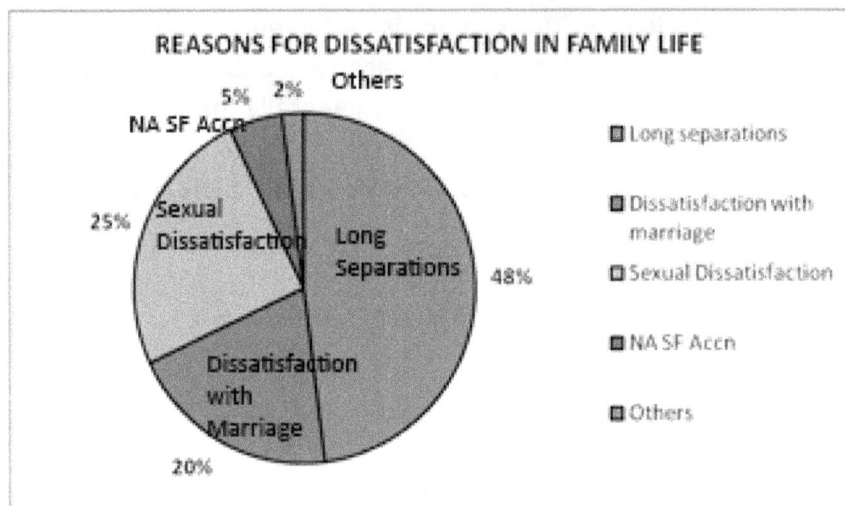

- Long separations
- Dissatisfaction with marriage
- Sexual Dissatisfaction
- NA SF Accn
- Others

ORGANISATIONAL REASONS FOR SUICIDES

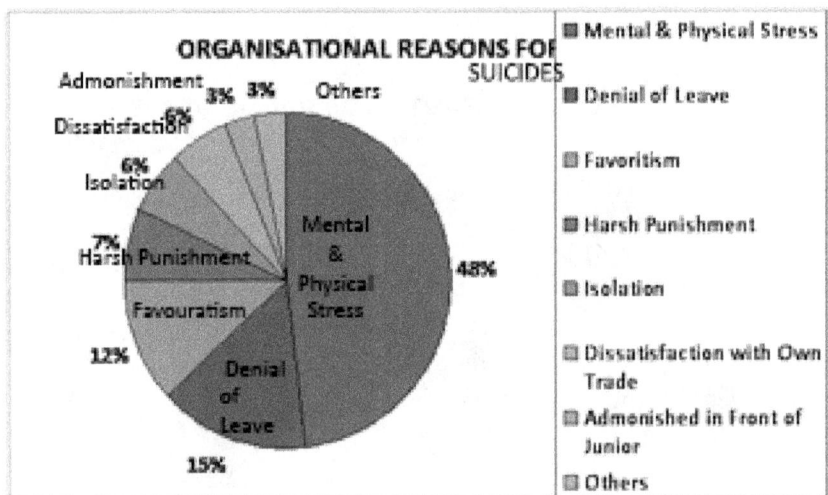

- Mental & Physical Stress
- Denial of Leave
- Favoritism
- Harsh Punishment
- Isolation
- Dissatisfaction with Own Trade
- Admonished in Front of Junior
- Others

Self Inflicted Injury

Self inflicted Injury is generally associated with malingering (an attempt to avoid hard life or difficult operations). However, it cannot be ignored as it could well be a case of attempted suicide. Without going into any specific data, it can easily be said that it is not a major problem for worry either in J&K or the NE Region. For example in J&K, there is one case of Self Inflicted Injury for every six cases of suicides. And in the NE Region, there in one such case for every eight cases of suicides.

Conclusion

While the army focuses on the non-systemic issues and improves the environment in its combat units, the government must take up the systematic issues with urgency. The aim must be to eliminate the cases of suicides. At the same time, there is an urgent need to place the army and indeed the armed forces at levels which restore their "*Izzat*" and **honour.**

There can hardly be any doubt about the reasons for the increasing number of incidents of suicides. Many of us, as practitioners, of the art of soldiering (Both retired and serving) understand this fully well. The team of the psychologists, in their report (DIPR report), submitted to the Ministry of Defence, have only reminded us by adequately analysing the problem. This problem, however, needs more to be addressed than has been set out in the public domain. The advise rendered by the union defence minister to the Indian army calling for greater interaction between the officers and the men is fine but it must be realised at all levels that useful remedies will not be forthcoming unless some of the root causes of the problem are identified and addressed. These measures go beyond the conventional and stereotype methods. The army, on one hand, prefers to sweep some of these under the carpet, hoping that these will get resolved with the passage of time, on the other hand, there are actions/ decisions well beyond the realm of the army and need the involvement of not only the MOD but the nation as a whole to resolve them. While the army needs to face up to some of the realities after due introspection, deal with the inadequacies without any further loss of time, the responsibility of the political leadership and society towards the man in uniform cannot be wished away. Therefore, while rendering advice to the army, it is imperative that the political leadership and the civilian bureaucracy apply themselves in all seriousness to the systemic shortcomings and play a role in restoring to the soldier, a sense of dignity and pride in his profession.

Soldiers join the army to fight for their country and kill the enemies. They have no desire to kill themselves or their comrades, least of all their officers. Let us not make them do so, by a callous approach and not ameliorating their concerns. *Pious platitudes are no substitute for substantive action.*

CHAPTER 6 -FRATRICIDES IN THE ARMY

"A soldier raises his hand to salute his Officers and not to shoot at them"

The issue of Fratricide like Suicide is a cause of concern for the Army. There were negligent/ isolated cases of Fratricides even earlier but with the additional commitments in J&K, the number of such cases has gone up. This probably is the main reason for the rising incidents of Fratricides.

This chapter analyses the magnitude of the problem and addresses the causes of this malady.

Understanding Fratricides [29]

Fratricide is the action of an individual killing a member of his fraternity. The connotation goes beyond the term "Running Amok". It includes cases of killings through accidental firing (An aircraft dropping bombs on own troops or inadvertent killing due to heavy volume of fire). It also includes a member of the team in an operation and then attributing the same as a victim of enemy action.

Analyses of Fratricides. Each individual irrespective of the rank, cherishes his or her IZZAT (Respect); Sensitivity-caste, Religion and Belief and expects justice. The individual interacts with his superiors, Colleagues and subordinates in his day to day work. **Inter Personal**

---

[29] Extracts on Fratricides and Fragging from Wikipedia, the Free Encyclopedia

Relations therefore, cordial or otherwise play an important role. During the performance of his/ her duties, mistakes intentional or otherwise are bound to occur. He/She is therefore likely to be taken to task and the senior could lose his cool. In the process, the senior may not only use intemperate language but could also punish. If such language/punishment is used or meted out in public, it would mean insult or humiliation of the individual. Else the senior may cast aspersion on the individual's caste, religion, belief or even family. There may also be instances like grant of promotion, detailment on duty or grant of leave where the individual feels that he/she has not got justice. In all such instances, an individual is likely to be provoked to the extent of using a weapon to kill the person who has provoked. When under stress or under the influence of alcohol, even a simple argument could spin into a fight, leading to shooting.

From the above analyses, it is evident that Fratricide is an inter play of two factors viz Sensitivity of a soldier and Interpersonal relations. This has been clearly brought out in the **"Report of the standing committee on Defence -08-09"**. According to the report, Fratricide is more of personal nature and does not have so much to do with the organisation. A person takes this drastic step when he is very much antagonistic towards that individual and has personal grievances against him. While this may not be entirely true but it does highlight the importance of interpersonal relations and sensitivity. It would therefore be prudent to understand the various dimensions of these two important facts.

Inter Personal Relations [30]

There are different perceptions and views with regard to Officer-Man Relationship. Some feel that this relationship is as good or as bad as it used to be. It is only that the environment has changed. The officers and men are not the same. The men are better educated and their expectations/ aspirations of how they should be treated

---

[30]   Extracts of Report conducted by 92 Base Hospital.

have gone up. They expect more courtesy and civility when being addressed. The officers, on the other hand, have hardly any time for their men. Being under pressure, they tend to be abrasive. Their perceptions have also changed. That is why some senior retired Army officers are concerned about the changing equations between not only officers and men but also those within the officer cadre: the senior officers and their subordinates.

For a clear understanding of this factor, there cannot be a better reflection on this issue than what has been done through a study conducted by 92 Base Hospital, under the aegis of the then Director General, Armed Forces Medical Services. Details are contained in **Appendix F** Attached.

Sensitivity of the Soldier

Salient points are contained in succeeding paragraphs.

There are different opinions with regards to the sensitivities of a soldier. Some believe that a soldier is a soldier irrespective of which country or Army he belongs to. All soldiers respond similarly to a given situation or are equally affected by the same external pressures. All soldiers are upset by criticism, humiliation or harassment.

There is however another school of thought that soldiers of all groups, races, nationalities and cultural heritage cannot be grouped under common susceptibilities and responses. Differences in levels of education, culture, value systems, material well being, social mores, responses and psyches exist. These differences influence the reactions of different cultural and national backgrounds. Similarly motivation factors not only differ from nation to nation but generation to generation. Accordingly, officers, junior commissioned officers and men develop distinct personalities.

The Emotional Needs of the Indian Soldiers The average Indian soldier belongs to the poorer section of the society mainly with rural background, less education and less worldly exposure. Consequently their material needs are minimal. Nevertheless their emotional linkages with their families are high. Therefore, despite high conditioning to the requirement of the service, emotional handling constitutes the critical factor. During prolonged separation from their families and children due to protracted war, they find it difficult to manage their finances, especially in view of low pay scales. Other domestic problems that confront him are the education of his children, security and management of small holdings and farms which may be in their family possession. Often family feuds at home become the major disturbing factor for which they may leave at short notice. On such occasions, they may require not only some additional money but also some counseling. Under LIC conditions, it may become difficult to grant them frequent leave.

Uncertainty of money and inability to be with his family when required become high triggers of discontentment. The liberal facilities of communications through STD calls, possession of Mobiles and even a letter do help in either diffusing or even aggravating their mental tensions, due to their inability to do their duties towards their families, and dependents. The problem is further compounded in case the wife of the jawan has a mobile with her. Having a sympathetic attitude, doing justice and showing affection along with superior administrative skills thus remain the sheet anchor in holding the troops in the right frame of mind and function.

Lt Gen SC Sardespande in his article on "Psychological and Motivation factors affecting the soldiers published in the Combat Journal, Apr 88 writes:-[31]

- Patriotism was too distinct a dream for our jawans, regimental spirit is the life sap and Leadership is the solid propeller.

---

[31] Psychological and Motivation Factors affecting the soldiers by Lt Gen S C Sardeshpande- Combat Journal, Apr 1988.

- Creature comforts and even welfare measures are a substantive replacement of the hard, tough and long tenures on our borders.

- Fear, Anger, personal harm and injury have lesser effect on officers but more on JCOs and men.

- Defeat and disgrace are dreaded by all. Unit disgrace is more dreaded than personal by all except the JCOs.

- Bad discipline because of bad living conditions figured very low. (*This no longer holds good because the men of today seek more comforts now*). Most of the JCOs and men attributed ill discipline due to lack of welfare, prolonged tenures and bad leaders.

- Soldiers get violent because their colleagues and officers don't solve their problems.

- Religion and sentiments play very little role as a cause for anger and violence.

- Commanders, GOD and Prayer are the main props to seek solace. Main cause of fear is the lack of knowledge and Commanders briefing. NCOs and soldiers put hunger and fatigue next. Discipline, tough training and tough Commanders are an antidote for fear.

- Good weapons, good training and good leaders are favourites for high morale.

According to the survey carried out, he concludes that despite hardships of terrain and tenure, our officers and men are reliable. They are generally unaffected, simple and pragmatic. Despite some phlegmatic soldiers, the Army as a whole is alert, kicking and ticking.

Not withstanding the fact that our Army is in fine shape, there are some aberrations- the behaviour of some our soldiers has led to the series of Suicide and Fratricide cases. Apparently the soldier has become more sensitive to certain aspects like his family, religion, caste and so on due to certain external impressions and influences that have altered his behaviour.

Now let us have a look at the factual data and relate the relevance of the above.

Fact Data Sheet

Refer to Appendix G attached.

Evaluation of Data

The evaluation of the data reveals the following:-

- The rates of fratricide cases, though more in the infantry battalions, are less than the RR on pro rata basis.

- Majority of the cases of fratricide are in the age groups of 20- 30 years.

- Fratricide cases pertaining to troops from the rural area are almost three times to those hailing from urban areas.

- Majority of the fratricide cases are done/committed by the matriculates.

- Troops from the Northern region top the fratricide cases followed by those from the Eastern region.

- Maximum cases of fratricide are against the NCO's/OR's followed by JCOs and the least number are against the Officers. (Shows that the soldiers still carry more respect for their officers).

- Maximum soldiers kill themselves after shooting their officers followed by the JCO's. (This again reflects remorse on the part of the soldiers when they shoot their Officers/ JCO's.

Else it could be the fear of consequent punishment. The former seems more likely).

Provocation. Provocation is precipitated mainly by humiliation, criticism, family problem and non grant of leave, Use of Abusive/ Intemperate language, unjust distribution of duties or under influence of alcohol. Out of these humiliation is the biggest provocation (Refer to Article, Fragging : Humiliation, biggest trigger- *Times of India, 09 Jun 2007.*

Studies/Reports

The following studies/ Reports relevant to Fratricides have been referred to:-

- Extracts from the study by 92 Base Hospital (Appendix F)

- DIPR Report.

- Article on Fratricides in the Army. [32]

- Study extracts from NDC Dissertation.

Evaluation of Studies

Study extracts from NDC Dissertation.

- No. 1Cause - Sense of feeling aggrieved & humiliated.

- No. 2 Cause - Inadaptability to Army Life.

- No. 3 Cause - Lack of Motivation and Pride in the Unit.

- No. 4 Cause - Punishments in public/ Man Handling.

---

[32] Fratricide Incidents in the Army by R S N Singh-Indian Defence Review, 2004

- No . 5 Cause   -   Use of Abusive/ Intemperate Language.

DIPR Report. Survey in both the Commands (Northern and Eastern) indicates the following causes:-

- **Officers.** Believe that lack of cordial relations between the leader and the led is the cause.

- **Jawans.** Believe that Humiliation/ Harassment by officer higher up is a significant trigger.

Inter action with the troops in J&K reveals the following causes:-

- No. 1 Cause   -   Put down in front of others.

- No. 2 Cause   -   Non grant of timely leave.

- No. 3 Cause   -   Unjust distribution of duties.

- No. 4 Cause   -   Other reasons.

- No. 5 Cause   -   Work Pressure.

Causes of Fratricide.[33] (Soldiers running Amok)

Amongst the younger age group, unjust distribution of duties seems to be the chief cause of fratricide. The problem understandably decreases as the age increases. However, the seniors are intolerant towards being insulted in front of their juniors. Leave problem is another cause for fratricide.

---

[33] Data based on discussions/Interaction during visit to the valley

Causes of Fratricides

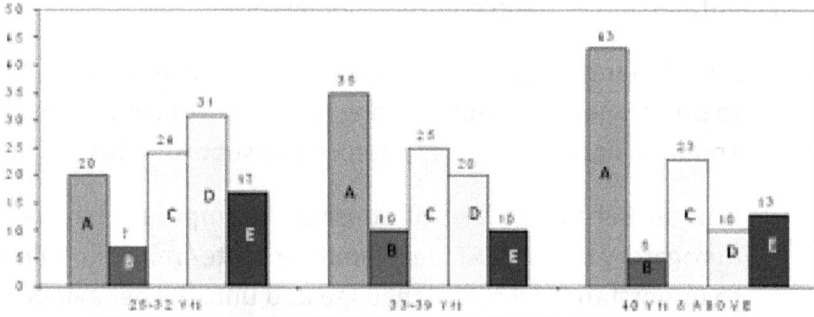

Legend:

A   Put down in Front of Juniors (Humiliation)
B   Work Pressure
C   Leave Problems
D   Unjust Distribution of Duties
E   Any Other Reason

Table3-Causative Factors Of Fratricides

Cause	25-32 Yrs	33-39 Yrs	40 Yrs and above	Remarks
Put down in front of Juniors (Humiliation)	20	35	43	No-1 cause
Work Pressure	07	10	5	
Leave Problem	24	25	23	No-2 cause
Unjust distr of duties	31	20	10	No-3 cause
Any other reason	17	10	13	

For soldiers running amok, the following sensitivities have been identified as the main contributory factors:-

- Punishments, the soldier feels aggrieved and humiliated due to punishments in public, wrong / unauthorised punishment and humiliation, especially before the subordinates.

- Human Behaviour. Use of abusive/ intemperate language derogatory to his IZZAT, ego, family or caste/ religion. Failure to understand the body language and unusual behaviour of troops, not granting promotion when due, sodomy, not granting leave especially in emergency and under the influence of liquor.

It is interesting to note that none of the causes of Suicides or Fratricides relate to hardship in difficult terrain, long tenures in LIC, inadequate living conditions or quality of food.

The Indian soldier can brave a lot of hardship under harsh/ difficult terrain conditions, can survive with makeshift arrangements for shelter and little food but he cannot withstand:-

- Personal insult, humiliation or harassment.

- Abusive language.

- Insensitivity to his family, caste/religion and domestic/ financial problems.

Most importantly, it must be understood that the soldiers are already under pressure situation. They therefore are in a state of anxiety, and frustration which leads to tension/Stress. When under stress the reactions are more emotional rather than rational. This understanding of the soldiers' state of mind in relation to his sensitivities will help officers to respond correctly and avert any untoward incident. They need to be cool. In fact any insult / humiliation with anger is more damaging and invites instant reaction.

According to the analysis by the author, the following are the main causes of Fratricides:-

- No. 1 Cause - Humiliation (Punishment in Public or Use of Abusive/ Intemperate language).

- No. 2 Cause - Lack of Sensitivity

- No. 3 Cause - Injustice(Unjust distribution of duties)

- No. 4 Cause - Non grant of leave in Emergency

- No. 5 Cause - Argument spinning into a fight

Accidental Firing/ Negligent Discharge of Weapons(NDW)

This chapter would be incomplete if there is no mention of NDW, an attendant problem of Fratricide as these cases could be an attempt to kill a comrade of one's own fraternity. These cases should therefore be taken seriously as cases of attempted Fratricide. There is little worry for the NE region as there were only two cases during six years. However, there is a cause for concern in J&K particularly as this phenomenon picked up only from 2006 onwards. Since then, there have been more cases of NDW than the cases of Fratricides.

Conclusion

The cases of Fratricides are generally a reflection of poor Inter Personal Relationship. It is also indicative of lack of sensitivity towards the sensitivities of the soldiers. If the relationship amongst various ranks is good, there is justice on the part of the leaders and every one is conscious of not hurting each other's personal feelings, then one can ensure that there will be no cases of Fratricides. The cases of NDW are an adverse reflection on the drills to be adhered to in the unit/subunits. It is due to violation of basic drills that leads to the cases of NDW, which cannot be ignored . Strict action needs to be taken for such cases.

Appendix F

(Refers to Chapter 6, Page 95)

EXTRACTS OF STUDY ON INTERPERSONAL RELATIONS AMONG UNITS DEPLOYED IN CI OPERATIONS

1.   Some of the major findings of the study on Inter Personal Relations and also related to stressful situations are enumerated below:-

- Jawans were found to be higher on neuroticism as compared to JCO's and Offrs. They were highly sensitive to stressful environmental situations and displayed low degree of stress tolerance. They also displayed low self esteem, excessive and conflicting motivations, emotional instability and inadequate coping procedures.

- Jawans who experienced most stressful conditions in insurgency areas, felt that the harsh nature of job was also an important cause of Inter-personal violence.

- An overwhelming number of Jawans, JCO's and Offrs have cited poor man-management, feeling of injustice, humiliation by seniors and domestic problems as the key causes of inter-personal violence.

- In an environment of constant threat to life, loneliness, lack of communication and inadequate inter-personal relations led to alcoholism amongst JCOs and Jawans who perceived these factors as contributing significantly to inter- personal violence. However Offrs did not perceive inadequate inter-personal relations and alcoholism as an important factor leading to inter-personal violence.

- Jawans and JCO's are found to be significantly higher on measures of pessimism as compared to Offrs. They had the tendency to look upon the future with uncertainty, disbelief and disdain.

- Jawans and JCOs were found to be more introverts as compared to Offrs. They were more self oriented, sentimental and idealist than realist. They appeared to be highly egoistical, involved in competitive situations and thus vulnerable to threat of failure.

- Empathy was considered important for people in leadership positions. Offrs deployed in insurgency areas perceived themselves to be more understanding to others' feeling, needs and sufferings as compared to JCOs and Jawans.

- The expression of anger differed drastically among Offrs, JCOs and Jawans. The Offrs preferred to give vent to their anger verbally i.e. by shouting, abusing, and humiliating etc whereas Jawans tend to become physical and thus resorting to beating and taking the extreme step of shooting.

- Among the Jawans and JCOs, the expression of anger was more displaced and intropunitive as compared to Offrs. The JCOs and Jawans displaced their anger on less powerful instead of the direct source of anger. Most of the time, it was directed inwards i.e. on ones self. The displaced and intropunitive form of expression of anger led to accumulation of frustration and feelings of helplessness. This tendency often resulted in Jawans taking up arms for seemingly small issues such as sanctioning of leave etc.

2. Overwhelming majority of Offrs, JCOs reported that poor man-management, injustice, domestic problems and humiliation were main causes of violence.

Table 4: Causes of Inter Personal Violence

**PERCENTAGE AND CHI-SQUARE VALUES OF FACTORS PERCIEVED/
NOT PERCIEVED (YES/NO) AS CAUSES OF INTER PERSONAL
VIOLENCE AMONG OFFICERS (87), JCOs (98) AND JAWANS (172).**

Causes of Inter Personal Violence	Officers			JCO's			Jawans		
	Yes	No	(x2)	Yes	No	(x2)	Yes	No	(x2)
	(%)	(%)	(%)	(%)	(%)	(%)	(%)	(%)	(%)
Poor Man-Management	85.5	14.5	29.82**	76.5	23.5	21.78**	80	20	80.01**
Injustice	79	21	19.76**	81.8	18.5	30.86**	85.3	14.07	83.3**
Domestic Problem	67.7	32.3	7.12**	72.8	27.2	16.00**	78.8	21.2	55.35**
Disrespect to Religion	54.8	45.2	0.40	49.4	50.6	0.00	52.3	47.7	0.29
Humiliation	85.5	14.5	29.82**	61.1	30.9	11.11**	74.4	25.3	40.52**
Lack of Communication	27.4	72.6	11.76**	61.7	38.3	4.00	67.7	32.3	20.48**
Alcohol	24.2	75.8	15.5**	69.1	30.9	11.11**	58.8	41.2	4.95*
Loneliness	40.3	59.0-7	1.95	70.4	29.6	12.64**	11.8	97.89	
Poor Interpersonal Relations	53.2	46.8	0.15	65.4	34.6	7.11**	74.7	25.3	40.52**
Nature of Job	46.8	53.2	0.15	56.8	43.2	1.23	74.1	25.9	38.59**
**p < .01 *p < .05									

Appendix G

(Refers to chapter 6 Page 94)

FACT DATA SHEET

Table 5:   Data: Fratricides (2002-JUL 2007)

(Data extracted from Standing Committee for Defence-08-09)

Year	No of Incidents	Total Casualties	Remarks
2002	17	21	
2003	05	08	Average cas per year=14
2004	05	11	
2005	06	13	
2006	13	24	
2007	07	08	
2008	0	0	

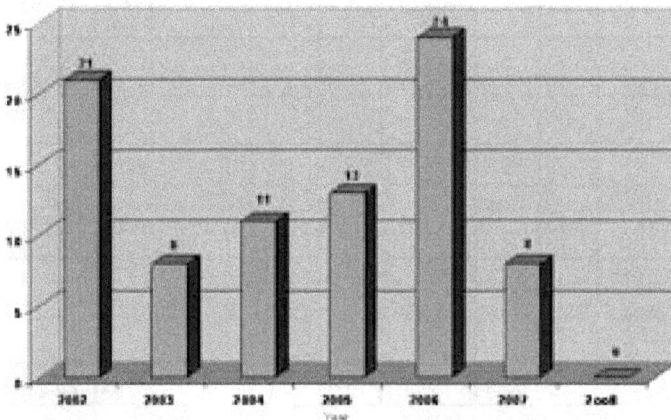

FRATRICIDE CASUALTIES

Table 6- Breakdown Of Fratricide Cases

Unit wise	2002	2003	2004	2005	2006	2007 (Till Jul)	Total	%
Infantry Battalion	09	01	02	01	04	03	20	40
RR Battalion	02	02	-	02	05	-	11	22
Others	04	02	03	03	04	03	19	38
Total	15	05	05	06	13	06	50	
Age Wise								
>20Years	03	01	02	01	03	-	10	20
20-25	07	03	02	03	03	01	19	38
25-30 Years	03	01	01	02	05	05	17	34
<30 Years	02	-	-	-	02	-	04	08
Total	15	05	05	06	13	06	50	

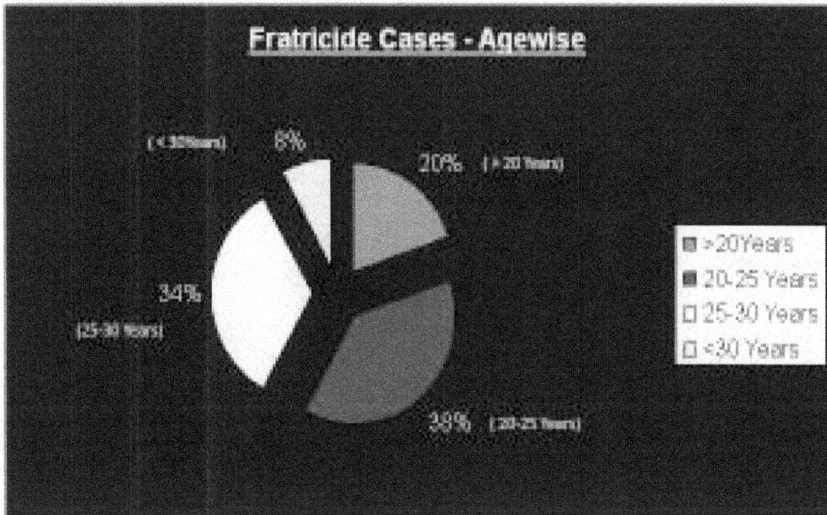

	2002	2003	2004	2005	2006	2007 (Till Jul)	Total	%
Edn Qualification								
Matric	15	05	05	06	13	06	50	100
Urban/Rural								
Urban	03	01	01	01	04	02	12	24
Rural	12	04	04	05	09	04	38	76
Total	15	05	05	06	13	06	50	
Region Wise								
Northern	03	01	01	02	04	03	14	28
Eastern	04	01	-	01	02	-	08	16
Southern	04	-	01	02	02	02	11	22
Central	02	02	01	-	03	-	08	16
Western	02	01	02	01	02	01	09	18
Total	15	05	05	06	03	06	50	

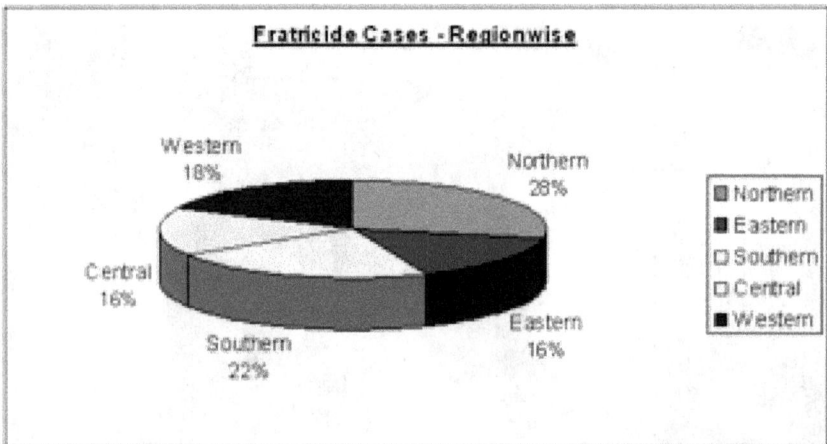

Fratricide Cases - Regionwise

Western 18%
Northern 28%
Central 16%
Southern 22%
Eastern 16%

Northern
Eastern
Southern
Central
Western

	2002	2003	2004	2005	2006	2007 (Till Jul)	Total	%
Rank Wise								
Against Offrs	03	-	01	01	03	01	09	18
Against JCO's	03	-	03	02	02	01	11	22
Against NCO's/- OR's	O9	05	01	03	08	04	30	60
Maximum cases against NCO's/OR's								
Cases where shooter shoots himself								
*After shooting Offrs	02	-	01	01	01	-	05	10
After shooting JCO's	01	-	-	=	01	01	03	6
After shooting NCO's/- OR's	01	-	-	01	02	-	04	8

CHAPTER 7: HUMAN COST OF ADVERSE PSYCHOLOGICAL EFFECTS OF LIC OPEARTIONS- IMPACT ON THE ARMY

"One obvious and tragic price of war is the toll of death and destruction. But there is an additional cost, a psychological cost borne by the survivors of combat and a full understanding of this cost has been too long repressed by a legacy of self deception and intentional misrepresentation. After peeling away the legacy of lies that has perpetuated and glorified warfare, there is no escaping the conclusion that combat and the killing that lies at the heart of combat is an extraordinarily traumatic and psychologically costly endeavour that profoundly impacts all those who participate in it "

> - Lt Col Dave Grossman and Bruce K Siddle (Psychology of Combat)

Soldiers getting impacted by combat stress are historical but for the Indian soldier, it is a recent phenomenon. The Indian Army has been involved in LIC operations for over five decades). But there were no reports of combat stress, though some cases of violence, suicides and fratricide were reported. The number was insignificant, not demanding any special attention. It was only after the Army's involvement in J&K that media carried reports of suicides and fratricides. The turn around became apparent after 1996 when the foreign terrorists joined the struggle.

Consequently, several articles appeared in newspapers/ magazines and some books were also authored on the subject of stress. This was followed by a series of studies undertaken by the army psychologists and psychiatrists, which gave some idea of the degree to which Indian army had been afflicted with the adverse psychological affects of LIC. But none of the studies brought out the impact on the Army in terms of Financial and Human cost.

This chapter deals with the human cost that the Army is incurring every year. From this, the financial can be worked out.

Studies

The following studies have been referred to:-

- Evolving medical strategies for low intensity conflicts- A necessity- (Lt Col Ajay Dheer et al).

- Psychological effects of LIC operations (Lt Col S Chaudhary et al).

- The Indian Army: The Crises within, India Together,13 Sep,2007.

Evolving medical strategies for low intensity conflicts- A necessity- (Lt Col Ajay Dheer et al). Common Symptoms experience by troops is given in Table 7 below:-

Table 7- Common Symptoms Experienced by Troops (Lt Col Ajay Dheer et al)

Ser No	Indicator	Likelihood (%)	Remarks
1.	Tiredness	100	Not satisfactory
2.	Nervousness and Tension	95	
3.	Sleep Disturbance	95	
4.	Hyper Arousal	90	
5.	Anger & Irritability	85	
6.	Loss of Interest in work	70	
7.	Feeling of being victimized	62	
8.	Disobedience/Hesitation to work on orders	65	
9.	Loss of Appetite	62	
10	Pessimism and Sense of Helplessness	55	
11.	Disorientations when faced with a changing situation	45	-
12.	Sluggish response to orders/Changing Situations	42	-
13.	Urge for Alcohol	35	-

(Study by Ajay Dheer et al- Medical Source-Studies for LIC)

Alcoholic abuse. Alcoholism is one of the major problems in the Army. Research carried out on the subject has this to reveal-[34]

- India is a low alcoholic consuming country (0.94% of absolute Alcohol).

- Consumption increased recently (2003 onwards) by 106% vis-a vis other countries.

- 50% cases of those discharged on medical grounds constitute alcoholic cases.

- In the Army, out of 245 cases of alcoholic abuse, rank wise details are- (Offrs-3.74%, JCO's-26.9% and OR's-65.7%). Maximum cases are from the DSC.

- Field Area-48.69% and Peace Stations-51.31%

Neurosis. According to a study, the following is revealed-[35]

- Overall prevalence-31.34% with 95% confidence interval; Personnel-32.59% and Families-30%

- Gender and Marital Status had no effect on prevention of Neurosis.

- Soldiers in Age group; 25-36years age group is most affected.

- More common in lower ranks and among troops from Arms.

- Among Wives- Age, Rank and type of Arm/Service did not affect.

---

[34] Study on Changing Pattern of Alcohol Abuse in the Army before and after AO 3 & 11/2001 by Col S Saldhana et al

[35] Study on Neurosis among Armed Forces Personnel by Maj Dheer et al

Major conclusions on the impact of Adverse Psychological Effects of LIC as per the Study Report by S Chaudhary et al are contained in Appendix H.

Overall figures of affected persons are given in table below [36]

Table 8- Over all figures of affected persons (The Tribune, Saturday, Jan 6-2007)

Ser No.	Year	Psychological admissions in MH	No. of persons boarded out
1.	2000	2709	457
2.	2001	2783	345
3.	2002	4514	522
4.	2003	4432	238
5.	2004	4982	443
6.	2005	Not Known	401
7.	2006	Not Known	354

Note: - The important point is that while the overall percentage of cases in the Army may be negligible, the cases boarded out due to various reasons are significant ; Alcoholic cases-50%-Neurotic Cases—25% and Psycho Cases——25%

- Average casualties due to suicides per year= 100

- Average casualties due to Fratricides per year=9

Impact

The overall Human Cost due to Stress, Suicides, Fratricides and Mental Disorders is tabulated below:-

IMPACT OF STRESS

- Average cases of Mental Disorder
  (Anxiety, Depression etc) Medically
  Boarded Out per Year (Admissions - 4000)   -   401

- Average Suicides Per Year   -   100

- Average Fratricide Casualties Per Year   -   9

- Total   510

- Average Casualties in Encounters   -   359

- Human Cost in CI Operations   -   869

- Deaths due to Other Causes   -   1347

- Total Human Cost Per Year   -   2216
  (Two Infantry Battalions Strength)

A PERSON UNDER STRESS REACTS MORE
EMOTIONALLY THAN RATIONALLY

Conclusion

In conclusion, LIC imposes a psychological weight on the military mind, resulting in the feeling of stress and strain, generally described as being tense or being in a pressure situation. This results in depression, fatigue, alcohol abuse and psychiatric/ neurotic disorders. In extreme cases, this results in stress, suicides and fratricides.

The *overall impact* of human cost can be summed up as follows:-

- On an average, there are approximately 4000 admissions in the military hospitals in the form of cases of neurotic/ psychiatric disorders, out of which approximately 400 are medically boarded out.

- Fatal cases on account of suicides and fratricides are around 100 and 9 per year respectively.

- The total human cost per year on this account is approximately 510 persons.

Appendix H

(Refers Chapter 7, Page 113)

MAJOR CONCLUSIONS ON THE IMPACT OF ADVERSE PSYCHOLOGICAL EFFECTS OF LIC OPEARTIONS ON THE INDIAN SOLDIERS

1.   The soldiers showed high adaptability and inherent resilience. Better resilience is due to better cohesion of units, good leadership and high morale.

2.   They, however, had significantly higher levels of fatigue (general, mental and physical), depression, alcohol abuse and psychiatric disorders. These are areas of concern. Levels of fatigue were higher than that of the recruits in the training centres. Stress and anxiety levels are also not satisfactory.

3.   Psychiatric breakdowns remain one of the major areas of concern with distinct symptoms. This constitutes the maximum human loss when expressed in human terms. There has been a greater possibility of becoming a psychiatric casualty than being killed/ wounded. It is therefore, essential to comprehend the magnitude of the inevitable adverse psychological impact of LIC.

4.   Adverse psychological affects are related directly to the intensity and the length of service in LIC operations.

5.   Troops average about 16 hours a day in active operations with 5 hours of sleep and one off day in 11 days.

6.   There is no significant difference in satisfaction with life and personality traits.

7.   Troops are more frustrated in LIC, easily upset, tenser, feel more isolated and have decreased work efficiency.

8.   Some soldiers have problems with aggression, resulting in their running amok and shooting their colleagues or superiors on trivial matters.

9.   Despite the adversities and hardships of service in LIC areas, the morale of troops is high and general attitude towards life is optimistic. The integration/cohesion of units/subunits is intact.

10.   LIC operations are primarily young officers, JCOs and NCOs wars. *The top-down approach would not do.*

11.   The importance of indoctrination, training in public relations and strengthening of junior leadership cannot be overemphasised.

CHAPTER 8 : LEADER- LED RELATIONSHIP AND MORALE AND MOTIVATION

SECTION 1

LEADER- LED RELATIONSHIP

"Sensitive and caring leadership is the key to eliminate/minimise Suicides and Fratricides".

The soldier when under difficult conditions and situations seeks solace in his commanders as much as in God and prayers. History is replete with examples where leaders have motivated their men in achieving victories, against all odds. At the macro level, names of famous military leaders, like Field Marshal Rommel of the German Army, Montgomery, Slim and Auchinleck of the British Army and Patton of the US Army, are parts of a legend, for turning the tide of defeat into victory.

But in LIC, the operations are conducted at the platoon, company or at best, battalion levels. Therefore, there is a greater demand on the junior leaders, including JCOs and NCOs. The subject of officer –man relationship therefore should encompass all types of leaders down to the junior level. It will thus be appropriate to deal with the subject as Leader-Led relationship. And to start with, let us evaluate the type of leaders we have.

Leadership-An Appraisal

For the purpose of an appraisal, leadership has been divided into four groups as under:-

- Junior Leaders   -   (JCOs and NCOs)
- Junior Offrs   -   (Offers up to the Rank of Major)
- Middle Level Offrs   -   (Lt Col and Col)
- Senior/Higher Leadership -   (Brig and Above)

Junior Leaders. They constitute the back bone of the unit. However, there is a feeling among the Offrs that they do not shoulder enough responsibility. The JCOs in particular seem to be only acting as the medium for passage of orders from the top and requests/ complaints upwards. On the other hand, the JCOs have expressed their feelings of being bypassed and not taken into the loop of decision making. The brunt of executing tasks therefore appears to be on the NCOs and that of decision making on the Offrs. But the reality seems to be different. While exceptions are always there, majority of the JCOs seem to be fulfilling their responsibilities and acquitting themselves better than they did in the Indo -Pak wars of 1965 and 1971. The ratio of Offrs casualties in these wars was 5.9% while that of the JCOs was 3.5%. In CI operations, while the ratio of Offrs casualties is 6.5%, that of the JCOs is 8.15%. This may be understandable because most of the actions in CI operations are at platoon level. In fact the quality of JCOs and NCOs has shown a qualitative improvement after the introduction of the Junior Leader Courses. They, however are still prone to parochialism and favour those who may distantly be related to or belong to the same village/district. This manifests itself in the unjust distribution of duties and favouritism in grant of leave. The Jawans of today are sensitive to these and resent them much more than their predecessors did.

Officers. As regards the Offrs, the opinions are sharply divided. However, most of them feel that aberrations apart, the leadership of today is as caring and sensitive as their predecessors. For the purpose of appraisal, the leadership of Offrs is divided into three categories:-

> • **Junior Offrs (up to the rank of Maj).** In his article, "Officer-Man Relationship" [37], Lt Gen Satish Nambiar (Retd), ex Director at the USI, has said this about this lot of officers; "these officers have staked their lives, lost their limbs but never let the country or the Indian Armed Forces down". Many others like him have also commented on the commendable performance of the junior leadership; commencing with the Operations of 1948-49 in J&K, through the Sino- Indian Conflict of 1962, the Indo- Pak operations of 1965 and 1971 as also the counter insurgency operations in the North East since the 1960's and the anti terrorist operations in the Punjab, OP PAWAN in Sri Lanka and presently the on going operations in J&K for the last over two decades. In the Kargil operations of 1999, their brave performance converted an adverse situation into victory. The performance of the young officers has been outstanding by any standards of military professionalism. It is no longer a secret that the Kargil operations would never have been a success, without the determination displayed by the young officers and the rank and file of the Indian Armed Forces. It was their initiative, dedication to duty and the willingness to make the ultimate sacrifice of their life that gave us victory. Every body is unanimous about this lot of Offrs being daring and bold. They never stop taking risks and have exposed themselves to the same danger and risks as the men. They have done the country proud and have not hesitated in making the supreme sacrifice of their life when the situation so demanded. They have

---

[37] South East Asia Defence and Strategic Review Sep- 2007

delivered in all wars and conflicts. However, there is a growing opinion that they are not devoting enough time for the welfare of their troops and have neglected man-management. They have also been insensitive to the sensitivities of the men, resulting in poor inter personal relations. Some get away with an excuse that they do not get enough time due to the shortage of Officers at this level and the consequent pressure of work on the few that are present. While this is true but this is no reason to over look the vital aspect of man-management and good inter personal relations.

• **Middle Level Offrs (Lt Col and Cols).** They constitute the main planning and decision making body at the unit level. They also act as a link between the Senior/ Higher leadership and the subordinates. This lot of Officers is blamed for not acting as a cushion but seem to pass the pressure to the men as they seem to be having high aspirations. Not withstanding their aspirations in career, there has to be a balance between exerting the right pressure and flogging their command.

• **Senior / Higher Leadership (Brigs and Above).** The higher leadership needs to be more charismatic and humane. They set an example before the rest of the officers and men. But lately there is an increasing impression that some senior officers are conscious of hierarchy, status and privilege and observe double standards; one they preach and the other they practice, in person. While they preach austerity for others, they need luxurious standards for self. A few appear to be impatient and want only good results. In so doing, they tend to interfere with the command of the lower formations/ units. They also do not accept mistakes. When such mistakes occur, they take the unit to task instead of holding their hands.

Lastly, some of the greedy minds indulge in corrupt practices, never heard of earlier.

Not withstanding the opinions expressed above, on the whole, there is nothing drastically wrong with the officers. They are as good or as bad as their predecessors. But certain trends highlighted above are worrying.

Leader-Led Relationship

Now let us see what a soldier expects from his officer and what he thinks of him? This would be indicative of the relationship between the officers and the men. As regards the expectations, a soldier expects his officer to be robust; imbued with a strong sense of duty, honour and courage, professionally competent and concerned for his men. He would even stake his life to save his Sahib. He expects the officer to be different and gets disappointed if he is not. He expects the officer to set the right example at all times. To understand his impressions and thoughts about today's officer, one has to turn to frequent interaction one has had with those in service. There are some findings which are worrisome. The PBOR think that due to shortage of officers at the junior level, they pay insufficient attention to genuine care and concern for the subordinates.

The above perception of the officer class by their men shows that some of them tend to be self centred and uncaring for the soldiers. In sum, some of the officers of today do not treat the soldiers with due courtesy. They find little time for interaction and understand their problems. They, therefore tend to be insensitive to the sensitivities of the soldiers they command. —a cause for adverse inter- personal relations between the Leader and the Led, leading to cases of Fratricides.

The question therefore arises that if our middle and young officers are doing their operational job well, then why are there increasing cases of Suicides and Fratricides? Is the solution beyond them or do they have a role to play, as leaders, in reducing such cases? The answer is simply in the affirmative. But possibly there are some reasons beyond their control.

- Firstly there is a shortage of young officers. This has a serious impact at the operational level, especially in LIC where junior officers are required to under take operational tasks more often than they would have done for conventional operations. They are also required to handle additional command responsibilities to allow inescapable absence of their colleagues on other commitments like courses of instruction, temporary duty, and entitled leave. This problem is further compounded by deputation of the few that are available to the Rashtriya Rifles and Assam Rifles. Under such circumstances, the essential requirement of personal interaction with the men is left unaddressed.

- The nature of operations is also such that every one is under pressure and stress, the degree may vary. The officer is under greater pressure to perform. The general atmosphere is such that almost every one is charged up.

Therefore, while the officers are under extra burden due to the shortage of officers, they have no reason to lose their cool, in fact it becomes more important to control one's personal conduct and behaviour to avoid incidents like:-

- Meting out punishments which are unauthorised and asking the soldiers to undergo such punishments in public.

- Using abusive and intemperate language hurting the personal ego and IZZAT of the soldier.

- Being unfair in granting leave or promotions and so on.

That brings us to the larger issue of the style of command.

Style of Command

A proportion of the psychological stress borne by the leader and indeed imparted into the environment by the leadership is as a result of the dichotomy in the two styles of command prevalent in the Army. The two styles in question are the directive control and the Restrictive control. Directive control is delegating signifying trust in the hierarchy and enables exercise of the command functions at all levels. In contrast, the Restrictive control is the zealous hoarding of power and authority with the leader and an exercise of intrusive command philosophy by the leader. Of the two styles of command the Restrictive Control style is particularly dysfunctional in LIC environment. Where this is the preferred command style, the liberty of leaders at lower levels is constrained thereby increasing egregious stress levels in the hierarchy. Troops, comfortable with one style and subjected to the other, instead also require to adapt thereby increasing the strains they are already under, in adverse situations. They demand special leadership styles. LIC operations place onerous command responsibility on the leadership skills.

Leadership Requirements

The leader has to set an example, take personal interest in all aspects of the men especially their welfare and personal problems, talk to them, never lose temper, give credit where it is due, be prompt in awarding just punishment whenever it is required to enforce discipline, ensure physical fitness and be loyal. The aspects which merit special attention are:-

- Maintaining the sense of IZZAT and pride in the uniform by displaying an impeccable code of conduct, discipline and loyalty to the regiment ethos. A soldier is proud of his Regiment and will not do anything that hurts his IZZAT.

- Keep the distance but mix with the troops both formally while interviewing personnel going out or coming into the unit, Sainik sammelans, and informally at barakhana, games, religious gathering and so on.

- Make the channels of communication open, both formal and informal. Make the JCOs and NCOs accountable for reporting aberrations in the behaviour of any person. They must not be bypassed but involved in decision making.

- Should be able to handle a man in an agitated state in a calm and confident manner, even if he talks nonsense. Better sense would prevail soon.

- Keep the difficult under check by discreet surveillance.

- Set an example in austerity.

- High levels of Honesty and Integrity are vital in LIC environment in ensuring correct reporting.

- Men should be encouraged to put their grievances in Sainik Sammelans, interviews or any such forum. Their problems must be heard promptly and action taken.

- Ensure holding of Mandir and religious functions so that the religious teacher can give spiritual guidance.

- Develop the art of recognising stress symptoms.

- Look after their pay and allowances, publication of casualty, replacements of clothing, educational qualifications and recommend promotion, appointments, commission as per qualification and aptitude of the individual.

- Maintain the homogeneity of subunits as far as practicable.

- Correct selection of JCOs and NCOs. They are the backbone of command and control in units. Their prejudices, bias and rural background, responses need to be harnessed so that their religious and other sentiments of caste, creed and place of residence do not cloud their thinking and judgment.

- Combat stress can make troops ignore discipline .The soldier has a tough life. But then the Army is a profession of arms. There should therefore be no compromise on discipline. A jawan should be looked after but not pampered.

- Ensure courtesy and correct behaviour with the locals.

Conduct of Operations. The young and middle officers must carry out a detailed briefing of the impending operations. During the conduct, they must lead from the front. The officer must expose himself to the same danger as his men.

Welfare of Men. The men consider their officers as role models and look up to them to solve their domestic and personal problems. This would mean interacting with the district officials (DC and SSP). The officer must also ensure the men get good quality food. He must take care to provide three hot meals, cup of tea, cot and water for bathing. The officer must share the same hardship and improve the habitat so that the men get comfortable sleep and rest when not out on operations.

Problems Specific to Senior / Higher Commanders

The senior / higher commanders play a vital role in alleviating stress levels in the troops. They need to remember the following aspects :-

- **The Pressure of Numbers Game**. The phenomenon that has profound Psychological impact on the troops in LIC Operations is the body counts or "The Kills and Recoveries" number game and the negative competitive spirit amongst units deployed in CI Operations. All and sundry seem to shift the blame to the next commander for the turn of events which they feel is primarily due to the ambition of the commanders to achieve higher results. Where does it start? Without debating the issue whether the pressure is generated by higher commanders or by ambitious commanding officers, it must be emphasized that no one irrespective of the rank structure should be able to push the men below to achieve higher results. It is inherent in the task and needs no further goading for it.

- **Patience, Tolerance and Responsibility**. In CI Operations there is a need to understand the word "Patience and Tolerance". It is a patience game and senior commanders up to the top need to tolerate events even when they get out of hand. If all commanders want results, then they must be mature and tolerant enough to accept the reverses and failures too as these are the ground realties. However accolades are showered on the same company commanders and youngsters, when they get results (incidentally, they are the ones who get results). Then why is it that they are taken to task when, in their exuberance, they make mistakes? (What ever may be the cause). Such mistakes are seldom tolerated and not many care to hold their hands. No officer makes a mistake deliberately. Cases like collateral damage, killing of innocents or custodial deaths or any other HR violations are common fall outs of CI operations. On the contrary, the individual is not only left to fight his battle alone but also faces harassment/ fear of reprisal from his own organisation. This is where the basic definition of the word Responsibility needs to be recalled:-

  - When everything goes bad, I did it.

  - If anything goes some where good, then we did it.

  - When every thing goes good, you did it.

Interference in Command. There is a tendency on the part of the senior commanders to fight the battle of the commanders at the unit/ subunit level. This amounts to interference in command and needs to be avoided. While giving a free hand to the subordinate commanders, senior commanders should concentrate on the following aspects:-

- Gaining intelligence of strategic value.

- Managing the media.

- Managing the environment.

Keeping all the leadership requirements, discussed above, in view, it is time to evaluate the military leadership of today. What should be the role of the leaders at the middle/junior level? The answer is in three simple words: **Daring, Caring and Sensitive.**

- "**Daring** "means the leader must expose himself to the same danger and risk of life as his men.

- "**Caring**". It does not imply pampering the men but if one follow what Field Marshal William Slim singled out for his officers, for guidance, *"I tell you as officers that you will not eat, sleep, smoke, sit down or lie down unless your soldiers had a chance to do these things. If you will hold to this, they will follow you to the ends of the earth. If you do not, I will break you in front of your Regiment. While doing so, you must keep yourself separate from them and not try to be their buddy".*

- Being "**sensitive**" simply means being correct with your language and behaviour.(Refer to an Article, Sensitive Leadership key to arresting suicides by Girja Shankar Kaura-*The Tribune, 09 Jul 07.*)

As regards the senior officers, they must set an example, be charismatic and humane.

A Leader's Role in Creating Stress

It has been observed that the leaders tend to increase stress levels. This is how they do it:-

- Many leaders believe they must never display weakness or vulnerability.

- Above creates a situation in which leaders deny their own stress levels.

- Stress thus accumulates month by month and year by year. So leaders deny presence of stress beyond emotional, physical and relationship damage.

- Leaders need to struggle with these questions about their beliefs.

- Is the leader a lesser person if she /he acknowledges presence of stress in him?

- If one leader is to manage stress successfully, will that person impair performance or ambition?

- Can work be challenging yet free from stress?

- Does will power aggravate stress? If so, what is the alternative?

- How much personal investment is required to manage stress successfully?

In view of the above, every leader needs to evaluate himself and see how much he is responsible for stress and where he needs to take remedial measures.

Conclusion

Joining the Indian Army used to be every aspiring young professional soldier's dream. Field Marshall Montgomery was not selected for the Indian Army. However, he had to say this for the Indian soldier, "The Indian soldier responds to the leadership in a most remarkable way, and once you have won his heart, he will follow you any where". This remains true even to this day despite all the changes that have taken place. The Indian soldier is one of the best in the world, that one has the privilege to command. If we do not achieve our goals it is because of our short comings and not on account of the weakness of our men. Let us be reminded of an officers' prayer:-

"Lord make me worthy of the men I serve
Worthy of their loyalty and devotion to duty
Their wonder's willing and ready laughter.
Their great humility they ask for so little.
And give so much so readily without complaint
Grant their simple wishes Lord and bless them please.
For in this world no better soldier breathes than these."
(Greatest soldiers - song by Col C L Proud foot)

Section 2 - MORALE AND MOTIVATION

In LIC, a stage comes when the situation stabilises and a see saw battle of victory and defeat sets in. In such a situation, it is the will that matters, the will of the security forces to continue maintaining the pressure and the will of the militants to sustain it. The will of the troops is maintained through good morale and motivation.

Good morale and motivation is dependent on good man management and Officer-Man relationship. In both cases there are some issues which are either detrimental to good morale or those that reinforce morale. Various Research/ Studies have been published on it. These are:-

- Col D S Goel, Psychological Aspects of CI Operations.[38]

- Research Paper on Psychological Operations by Lt Col S Chaudhary (Retd) et al. This is similar to the study by Col Goel.

- DIPR Report.

Extracts of Col Goel's Study is given at **Appendix J**.

The findings and the conclusions of the above report is given in succeeding paragraphs.

Negative Detriments of Morale

The negative detriments of morale according to these studies / Research Papers are mainly relating to Operational and Domestic/ Social Issues.

- **Operational Issues.**

  - Fighting with constraints(one hand tied behind the back)

---

[38] Psychological Aspects of CI Operations- *Combat Journal, Apr 1998*

- Public Admonishment.

- Ambiguity regarding Aim.

- Feeling of Uncertainty.

- Ever Present Danger.

- Unable to deal with Jamayatis.

- Futility of war.

- **Domestic/social Issues**.

  - Inadequate Pay and Allowances

  - Hostile Population.

  - Disgust towards Corrupt Polity.

  - Concern for Family.

  - Lack of societal support.

  - Pangs of separation.

According to the Research by Lt Col Ajay Dheer et al, combat stress is caused because LIC is a campaign of nerves with less military and more Psychological with soldiers invariably fighting with hands behind the back. The strain of having to fight one's own people is infinitely greater than fighting the enemy. Environmental stressors which adversely affect a soldier's operating efficiency in LIC are poor living conditions and tenures extending beyond two years. He finds uncertainty accentuating the loneliness and monotony of soldier's life in almost everything; the very purpose of life, of living, of dying or being maimed, lack of information, the next contact with the militants, of what tomorrow holds for him, the duration of their deployment in a given situation, his role and his success etc.

Fighting with Constraints. In LIC operations, unlike a conventional war, the Army cannot use its full potential for fear of causing collateral damage. The soldier is also required to use minimum force and expected to capture rather than kill a militant. These constraints make a soldier feel that he is fighting with one hand tied behind his back..

Public Admonishment. The special moral values that a soldier is brought up with are often at cross purposes with those of an unconventional battle field. (Never a level playing field). Whereas a soldier is seen as a saviour in the conventional war, in LIC he is at the receiving end of public hostility. Unable to understand these conflicting reactions, the soldier is desensitised. The open aversion of the public to the security forces would affect anyone operating in such an environment because it is indeed humiliating and exasperating to see that the very people whom one is out to protect, turning against the force. Another aspect of public support is the lack of public recognition for some stupendous work done by the Army as part of civic action programme. It is not uncommon to see that while the public wants to enjoy the services rendered by the Army through these programmes, they remain hostile within. The hostility manifests itself through demonstrations, should there be a killing of an innocent man, even if advertently. Coupled with this, the hostile vernacular press projects the security forces as the perpetuators of oppression.

Ambiguity Regarding Aim. Unlike the conventional war where the aim is clear, in LIC there is some Ambiguity in the soldiers' mind as to why is he fighting this war. There is also ambiguity with regard to success i.e. are we moving forward or standing still?

Feeling of Uncertainty. In LIC operations there is always uncertainty in the soldier's mind regarding his being alive or dead, success or failure and lack of kills for a long time. Coupled with this is the lack of communication from the home front which makes him susceptible to his imagination, raising all sorts of thoughts, clouding his mind.

Ever Present Danger. The soldier is under constant fear of being attacked at any place or time and does not feel relaxed even in secure bases due to the danger of a suicide attack. Danger seems omnipresent, leaving the soldier tense.

Unable to deal with Jamayatis. The real problem is the Maulvis who preach hatred and motivate the young mind to become a militant. But the law forbids dealing with them. This brings only frustration in the minds of the soldiers.

Futility of war. When the soldier does not experience success he feels that he is fighting a futile war, which is unending. Complacency on the part of the political leadership to find a solution to the problem gives a feeling of aversion towards him. Sometimes even political interference in counter insurgency operations upsets the soldier as to why he is not being allowed to fight the battle his way. The worst is when amnesty is granted to the very militants who were captured, after launching several operations.

Domestic/Social Issues. Inadequate Pay and Allowances. The officers and men feel that they are not being compensated for the risk they are taking and the difficult conditions in which they are working.. There is an urgent need to de link the pay scales of the armed forces personnel from others and grant them better remunerations to make the armed forces an attractive career option.

- **Hostile Population**. The local population supports the militants, even if through coercion. They identify the Army as the killers of their kith and kin. They, therefore, while being sympathetic towards the militants are hostile to the army. This makes the gaining of intelligence about militants, difficult thus frustrating the soldier because it is like looking for a needle in the hay stack.

- **Disgust towards corrupt polity**. Insurgency requires a strong political will to resolve the problem. In most of the disturbed states this is lacking and conspicuous by its absence. Perforce the army has to shoulder the main burden of not only containing the militancy but also of carrying out development work since the civil administration is either defunct or unwilling/ reluctant to play its part of carrying out developmental work. Such apathy on the part of the Political leadership and the bureaucracy develops a sense of disgust in the mind of the soldiers.

• **Concern for the Family and pangs of separation**. The biggest concern for a soldier is the concern for his family back home. Particularly in case something happens to him. The soldier is constantly tense at the uncertainty of his family if he is killed in action. He misses them a lot as he is emotionally very attached to his family and thus feels the pangs of separation. The feeling that one is neglecting the family, can be a powerful stressor. The problem has been compounded with increasing number of nuclear families.

• **Lack of Societal Support**. The soldier of today does not hold a very high rung in the social order mainly due to financial in equality. This is true for all ranks, especially in our society which is fast adapting to global customs and practices, here the money is the mantra and knowing how to make it is knowledge. Our soldier is still steeped in the 19th century values and a British pedigree. Due to his lack of awareness, the soldier stands a dim chance of impressing the common man with the show of money, which has become the barometer of success in the society. He is generally regarded as one who lives in another world, bereft of the pulls and pressures of the modern society and visits the real world only when on annual leave, that too, for a brief period to experience the harsh realities. It will not be out of context to mention how soldiers are deprived of their small pieces of ancestral land, whose families are ill treated in joint families and whose widows have been cheated of the benefits they received on the demise of their husbands by their own kith and kin. If the family does not stay as part of the joint family, then she is harassed and at times even molested thus exploiting her being alone. The soldier is therefore a worried man in the absence of any support from others members of the family and the society at large. Add to this the empathy and lack of response from the civil administration in addressing the problems faced by a soldier in performing his duties in LIC environment.

Some of the other issues that have been covered in the book "Tackling Terrorism and Insurgency" by Maj Gen Samay Ram UYSM, AVSM, VSM (Retd) and contribute to combat stress, are given as under:

- Pressure from adverse media and Human Rights organisations.

- Intense Pressure of Combat conditions leading to lack of sleep and rest and exhaustion. Suffering own casualties.

- High expectations to get better kill results.

- Boredom and isolation.

Positive Detriments of Morale

These are self explanatory.

The Goel Study

Though the problems of combat related stress has received considerable attention in the West, particularly the United States, the Indian scene is, however, surprisingly sterile. While a number of authors have focussed on the politico-military aspects of such conflicts in the Indian context, few have touched upon the psychological issues involved. The first and perhaps one of the few formal psychiatric studies of the problem was carried out in the Northern Command, and the results have been reported. Then a study was carried out during Jun-Jul 97 by Col DS Goel, a leading psychiatrist of the Indian Army, among troops belonging to specialised battalions, engaged in high intensity CI Ops.

While a major part of the said study remains unavailable, some of the general findings have been incorporated in this essay. The material and method employed to acquire data for the study was a multidimensional approach which included:-

- Self administrated questionnaire consisting 83 structured and nine subject generated items, administered to officers serving in these units.

- Observer assisted questionnaire consisting of 42 structured and four subject generated items, administrated to Junior Commissioned Officers(JCOs) and Other Ranks(OR) serving in these units.

- Structured and semi-structured interviews with formation commanders at various levels within the theatre of operations and outside.

- Visits to forward areas to assess ground realities and to interact with troops deployed there.

- Complete psychological workup of individuals involved in aberrant acts with a view to build up a holistic profile of such incidents.

Confidentiality was assured by not asking for identifying personal details (Number, Name, Unit etc) from the respondents. Data was tabulated and subjected to computer analysis using a programme specially developed for the study.

Observations and Analysis. The response rate among officers was about 92.5%. There was a lack of equivocation among JCOs/OR with only 18.8% (approximately) giving "don't know" response. The demographic profile of the sample is reflected in **Table I.**

Determinants of Motivation/Morale. Not surprisingly a majority of the subjects chosen for the study were initially unhappy at being inducted into CI Ops, but their attitude became more positive as time went on. Motivational levels tended to decline when the tenure was prolonged more than 24 months. Most of the respondents felt that two years should be the optional duration of the duty in CI Ops. There were a small but significant number of personnel who had served for a total period of five years or even more in various CI Ops theatres. Some of the significant determinants of motivation and morale are as depicted in **Table II**. Contrary to general belief, leave was not a problem area and some of the main factors affecting morale are listed as under:-

- A feeling of insecurity with regard to families back home.

- Lack of societal support, adverse publicity in the media.

- Hostile attitude of Human Rights Groups.

- Lack of cooperation/hostility on the part of the local population.

- Dissatisfaction with regard to financial compensation with respect to additional risks/hardships involved in CI Ops.

- Difficulties encountered in rail travel while proceeding home on leave/posting.

- A sense of disgust towards a corrupt polity.

Operational Factors Affecting Morale. In the operational context, the factor exercising a negative impact included ambiguity with regard to aim, feelings of uncertainty, fear of ever present danger/attack from unexpected quarters, feeling of anger/ frustration at fighting with "one arm tied behind the back" and anger/bitterness at not being able to deal with the unarmed but vicious ideologues/ motivators/financiers of militants, the "jamayatis" who were blatantly misusing religious institutions such as 'Madrasas' in their anti-national activities. The same have been depicted in **Table III.**

Factors Reinforcing Morale. The above mentioned negative factors were counter- balanced by a widely held perception among the men that their officers were fair, just, and competent. There was good Regimental spirit and group cohesiveness. Notwithstanding the dangers and hardship involved in CI Operations as well as the, deficiencies in some areas, the feeling of organisational support contributed to high morale. The same have been depicted in **Table IV.**

Appendix J

(Refers to Chapter 8-Section 2, Page 132)

EXTRACT OF STUDY BY COL DS GOEL

TABLE I : DEMOGRAPHIC PROFILE

Rank	%of Total Sample	Average Age	Average length of Service	Average Length of Stay In CI Ops Area
* Officers	14.70%	34.68 yrs	15.14 yrs	15.53 months
* JCOs	18.20%	39.52 yrs	22.15 yrs	21.20 months
*OR	67.10%	28.25 yrs	10.70 yrs	18.70 months

TABLE II : DETERMINANTS OF MOTIVATION/MORALE

Parameter (Negative Determinants)	Officers	JCOs	OR	Average
* Feelings of insecurity regarding families back home	14.50%	42.20%	62.50%	33.80%
* Lack of social support	72.50%	38.60%	58.60%	54.50%
Adverse publicity in Media	52.40%	84.10%	30.20%	54.50%
*Hostile attitude of Human rights group	86.60%	24.40%	23.80%	44.20%
* Lack of cooperation/ hostility from local population	76.40%	68.50%	66.40%	78.70%

TABLE II : DETERMINANTS OF MOTIVATION/MORALE (Contd/-)

Parameter (Negative Determinants)	Officers	JCOs	OR	Average
* Financial dissatisfaction	90.20%	78.60%	89.40%	84.00%
* Difficulties in rail travel (Incl bribes demanded by railway staff)	40.40%	82.40%	91.00%	87.40%
* Disgust at a corrupt policy	91.50%	36.80%	61.40%	63.20%

Table III-Negative Operational Factors

	Officers	JCOs	OR	Average
* Ambiguity regarding aim	31.60%	28.60%	42.40%	34.20%
* Feeling of uncertainty	30.60%	40.40%	66.40%	45.10%
* Fear of ever present danger/unexpected attack	28.40%	42.40%	61.60%	44.10%
* Anger at fighting with constraints	76.40%	82.40%	84.50%	81.40%
* Bitterness at inability to deal with "Jamayatis"	91.50%	60.60%	66.40%	72.80.%

TABLE IV: REINFORCERS OF MORALE			
Positive factors	**Officers**	**JCOs**	**ORs**
* Regimental spirit	88.50%	78.20%	81.60%
* Group cohesiveness	89.00%	71.60%	84.50%
* Organizational support	91.00%	68.50%	72.40%
* Fear of letting down family prestige	31.40%	54.50%	42.40%
(Figures based on officer's assessment of what reinforces the morale of their men).			

Chapter 9—OTHER RELATED ISSUES

Section-1 - TENURES IN LIC ENVIRONMENT

"Protracted employment of the Army in CI/CT environment is
detrimental to its Morale "

The primary role of the Army is to defend the territorial integrity
of the borders of the country from external aggression. The secondary
role encompasses provision of aid to civil authorities when the law
and order situation goes beyond the control of the local Police and
the CPO Forces or when there is a major calamity like an earth quake,
floods or accident.

But for almost the last five decades out of 60 years of
independence, the Army was sucked into its secondary role. It started
with Nagaland and Mizoram in 1960's and thereafter in the rest of
the country, be it Assam, Punjab, or now J&K. What is important is
that the degree/duration of involvement increased and so also the
greater reliance of the State/ Central government on the Army. This
has resulted in almost a quarter of Army getting involved in CI
operations.

Today it has almost been accepted that the Army should continue
to remain deployed in LIC environment. The arguments are; firstly,
there is no other option for the government but to employ the Army
because the situation is well beyond the control of the Police/CPO

Forces due to foreign support to the militants and secondly, there has been no regular war since 1971. So, why should the Army not be deployed? Why should it remain a silent spectator to the threat of the militants?

Presently, therefore the Army is deployed both in J&K and the NE region to contain/ manage the situation and create a favourable environment for the government to function and start negotiations with the militants. This has become a reality, particularly in the context of the Police and CPO Forces being ill equipped and not fully trained. Who knows what happens in the Naxalites and Maoist affected areas. The involvement of the Army there too cannot be ruled out. The demand has already been made.

In view of the above, the lines between the Primary and the Secondary roles have got so blurred that the secondary role has become the primary role for the Army. Within the Army itself, there are strong advocates of the Army being trained for the unconventional war rather than the conventional war. But majority in the Army have resisted the continued deployment of the Army in CI operations. They, however, argue that it is adversely affecting the Army training and preparations for the regular war.

It was Gen Rodriques; the chief of the Army Staff in 1990s who hinted at the threat of over extension of the Army. He was referring to the constant deployment of the Army in CI/CT operations - intermittently ever since operation Wood Rose. This was followed by the war scare of Ex Brass Tacks and OP TRIDENT. This was closely followed by the deployment of over 75,000 strong Force of IPKF in Sri Lanka. De induction of the IPKF gave no respite; soon there after, the Army was deployed in Punjab in the West, OP RHINO and OP BAJRANG in Assam to restore normalcy along with its commitments that deepened in J&K and the NE region.

A similar voice was raised later by Gen VP Malik, Chief of the Army Staff in late 1990s. Even certain prominent defence analysts like Mr K Subramanyam voiced their deep concern over the increasing employment of the Army in the secondary role. (Also refer to an Article, Prolonged anti-insurgency taking toll on jawans by Rajat Pandit- *Times of India, 11 Jun, 2007*).

Not withstanding all the noise being made, the Army continues to be deployed for almost every thing from CI operations to disaster relief, restoration of law and order and mundane tasks like recovering of Prince from a deep pit. This tendency to employ the Army, much before the normal options available to the government are fully exhausted, is being viewed as not only the over use but abuse of the Indian Army.

The political leadership understands the negative implications of too frequent employment of the Army but probably has no alternative instruments to relieve the Army from CI operations. Regrettably, the CPO forces which have been raised for this purpose are neither well equipped nor well trained for this task. Numerous suggestions for upgrading the efficiency and professionalism of these forces have been muted but have met with bureaucratic stubbornness and turf protection. This has resulted in the Army getting increasingly involved in two widely separated regions, that is, J&K and the NE region. Therefore, no matter how much noise is made to limit the exposure of the Army, it will continue as the leading instrument for fighting Insurgency/ Militancy, in the foreseeable future.

The issue, therefore, is as to for how long a soldier should be exposed to CI operations so that he does not become a victim of combat stress that adversely impacts his mental and physical health.

Presently the units deployed in CI operations have tenure of two years. This chapter examines the viability of this tenure system and if not, then how can it be reduced further and by how much.

In this context it would be worth knowing what tenures have/ are being done by troops of other countries deployed under similar circumstances and conditions. **Table 9** below indicates the tenures of troops in Vietnam, Afghanistan and now Iraq:-

Table-9: Tenure of Troops of various countries in CI Operations

Country	Troops	Tenure
Vietnam	US Forces	18 Months
Afghanistan	USSR	12 Months
Iraq	US Forces	07-12Months* (Marines Only*)
Afghanistan	USAF and NATO	12 Months

The Indian Troops have tenure of minimum two years.

The first issue relates to the tenure of troops. Should it be one year or two years or more? The answer, if viewed from the point of exposure of troops to such an environment, is that it is better to opt for lesser tenures because lesser the exposure to violence, better it is for the mental health of the troops. Tenures of more than one year, tend to have adverse Psychological effects on the troops. But this is not entirely true because studies and statistics have shown that US troops in Vietnam, Soviet troops in Afghanistan and presently the US forces in Iraq and Afghanistan have had more cases of mental disorder than the Indian troops with two years tenure.

This raises another issue, that is, what factors decide the duration of tenures in such an environment? The answer lies in examination of some of the following factors:-

- Type of recruitment, Volunteer or conscription.

- The susceptibilities and responses of the soldiers from a particular heritage, race, nationality and culture.

- Degree of violence in the projected area of operations.

- The capability of a nation and its Army to affect turnovers after a given period, keeping in view the overall commitments of the Army, as also the financial burden of quick turn over of troops.

- The orientation period required and the utility of troops there after for optimum results. Troops can only be useful after spending a minimum of one year in the area of operations.

A detailed consideration of each of the above factors will indicate that a tenure of a minimum two years for the Indian Army is considered not only adequate but also useful and feasible to meet various parameters, without diluting the operational efficiency and putting too much pressure on the troops. This has been adequately supported by the findings of various studies on the subject. The main parameter is to however ensure that the total period of pre-induction training, induction, stay and de-induction does not exceed a total period of two years.

Random check of tenures of Infantry Battalions reveals that each unit undergoes a cycle of two years in LIC-Three years in Peace (A/ B Station)* - Two years in Type C station and then get inducted into a Field tenure of two years in CI/LC/HAA/ UN Mission. Thus every sixth year or so, an Infantry Battalion is back to CI/LC environment. Therefore, the problem is that even after two years of tenure in LIC environment and three years tenure in peace station, a unit generally gets tenure in either CI environment or HAA/deployment along the LOC.

This process of frequent induction, de-induction and induction again with consequent orientation and re- orientation leads to a sense of frustration and therefore stress in the minds of the troops.

Legend* :-

A   -   Very good Peace Station

B   -   Good Peace Station.

C   -   Average Peace Station

CI   -   Counter Insurgency Operations.

LC   -   Border Management

HAA   -   High Altitude Area.

UN   -   UN Mission Abroad

Conclusion

From the above discussions, it is apparent that the tenures of two years are manageable and have no major adverse psychological effects on the troops. However, the only issue is that some individuals, for various reasons like not being relieved on time or being earmarked again for tenure with R R, do longer tenures, resulting in serious problems. This must be avoided.

Section 2 - RASHTRIYA RIFLES- NEED FOR A RE- LOOK?

Introduction

The eruption of full fledged insurgency in J&K in 1990s placed a growing demand on the Army in dealing with this problem. Realising the increasing role of the Army to deal with LIC, it was felt that a force other than the Army was necessary, which could release the Army from this commitment. Thus the need of raising the Rashtriya Rifles (RR).

Since its raising, the force has performed well in contesting the LIC in J&K and has made a significant contribution in bringing the situation under control. But various problems that are inherent in the organization have compelled some of the senior Army Commanders to consider a re-look and recommend its re-organisation/ disbandment.

This chapter will examine the evolution of RR, its efficiency and impact on the soldiers and what needs to be done in the future.

Evolution of RR [39]

The idea of raising RR exclusively for fighting insurgency in J&K was conceived by late Gen BC Joshi, the then Chief of the Army Staff. During conventional war, the same force was to look after the Rear Area Security.

Accordingly, in 1990, the RR was to be established with two Sector Headquarters and Six battalions. The Army Headquarters in its case to the Ministry of Defence had made the following proposals:

- RR was to be a Para Military Force (PMF).

---

[39] The Rashtriya Rifles by B Bhattacharya- Bharat Rakshak Volume 3(2), Oct 2000.

- Its manpower was to be drawn from Ex Servicemen, volunteers from the Army and deputation from the Army.

- It was to operate /remain under control of the Army.

- It was to be financed by the Home Ministry.

- A review was to take place after three years.

The raising of RR ran into some problems:-

- Firstly no Ex Servicemen volunteered to join the RR.

- The Home Ministry declined to finance the RR.

The case was taken up with the then Prime Minister, Shri Narasimha Rao. He gave the go ahead for the raising and ruled that the case for funding could be resolved later. Thus, in 1994, the force came into being under Table 2 of the Army, (Under this Table, the strength is not counted towards the over all strength of the Army. The Infantry component comprised of 60%, Other Arms 25% and the remaining 15% were task oriented troops from various Services for logistics back up. The manpower was to be found by milking the Army units by 10- 20%. The Government also sanctioned additional 10 Headquarters and 30 battalions. Each RR battalion was to have six rifle companies with no heavy weapons of a battalion. They have better mobility with additional transport, better communication set up and logistics back up. Rs 600 crores were allocated to raise this force. In 1994, a formation headquarters equivalent of a Corps Headquarters was also sanctioned. This Headquarters was later bifurcated into two to form two operative Force Headquarters in J&K and a RR Directorate at Delhi. Within one year of its deployment, there were 30 suicide cases. This necessitated a review of the pattern of providing manpower.

Accordingly in 1995, a committee was constituted under Lt Gen Gurpreet Singh, the then Director General Infantry. The Committee found that there was lack of cohesion and integration in the RR battalions due to manpower coming from different battalions of the Infantry, and Regiments of Armoured Corps and Artillery. The committee therefore recommended that the RR battalions be raised under the aegis of different Infantry regiments and permanently affiliated to them. In 1996, a decision was taken to alter the composition of the RR battalions. Instead of its comprising troops from all over the Army, two to three RR battalions were made an integral part of each of the Infantry Regiments and other Arms. Their Commanding Officer also came from the same Regiment. But the component of other Arms/ Services remained the same. This ensured not only better cohesiveness functionally but also helped in developing the Regimental Spirit Corps.

One year left to review the organisation, there was no decision with regards to financing the RR. However, in 1995, it was decided to allot a separate budget to the Army, for RR only. The RR therefore remains a PMF under the Army. The Army provides the manpower and also the budget. In 1997, the Ministry of Home Affairs owed 950 Crores. In 1998-99, the budget outlay was 263 Crores. This was raised to 375 Crores. For the year 1999-2000, the outlay was raised to 587 Crores. By 2004, RR Battalions (1 to 57) had been raised and inducted in J&K. By 2005, RR battalions (58 to 66) were also raised thus completing the full complement of the RR Force (Five Force Headquarters, 17 Sector Headquarters and 66 RR Battalions for which in principle sanction was accorded in 2000).

Presently, the RR Force Headquarters are nominated as "Kilo Force, Victor Force, Delta Force, Romeo Forceand Uniform Force". Each Force Headquarters is headed by a Maj Gen. Their composition and deployment is shown in **Table 10** given below:-

Table-10: Location of RR Force Headquarters

- V Force   -   Anant Nag and Pulwama districts of South Kashmir.

- K Force   -   Kupwara and Baramulla districts in North Kashmir.

- Delta Force   -   Doda and Kishtwar districts.

- R Force   -   Rajouri- Poonch Districts.

- U Force   -   Udhampur district.

The Victor and Kilo Forces are under Headquarters 15 Corps and Delta, Romeo and Uniform Forces are under Headquarters 16 Corps.

Operational Performance

The initial RR battalions were deployed in the terrorist infested areas of Taran Tara in Punjab, and AnantNag in J&K. They proved to be extremely effective. In Punjab, the deployment of the Army and RR contributed considerably in the turn round in the situation. Since then, the RR is fighting the insurgency in J&K on behalf of the Army.

The supreme contribution made by the officers and men speaks for the contribution made by the RR. In 1991, 3 out of 44 casualties were of the RR. The figure went up to 8 out of 150 in 1996. Up to Feb 1997, 17 officers (including one Col, four Lt Cols and seven Maj), 13 JCOs' and 169 ORs' died. The ratio of officer casualty rates for RR at 1:9.94 till Feb'97 is double the average of the Army in Kargil War and 1971 war at 1:20.

With only 1/3 personnel changing every year and the permanent presence of the Sector and battalion headquarters, the RR battalions are most successful in creating intelligence network. This greatly helps in conducting CI operations.

By the time the RR celebrated its 8th anniversary, it had become the most decorated organisation of the Army. They had more than 600 Gallantry awards to their credit. 25% units were awarded unit citations by the Chief of the Army Staff for their distinguished performance up to this time. They had already accounted for 1516 terrorists killed. 3932 apprehended and another 458 surrendered.

Kargil Operations. During Kargil operations, a mountain division had to be pulled out of the valley. RR took over the extended responsibility. Besides, even some RR units took part in these operations. They were mainly employed in keeping the road and lines of communication open. While doing so, they took a lot of mine blasts. 26 RR battalions were deployed in the valley and three guarded the Banihal tunnel.

It was only after the Kargil operations that the government sanctioned Sector Headquarters Kilo and Romeo along with 20 more battalions.

Operational Assessment

The reputation of RR battalions as an anti-terrorist force has a tremendous impact on the militants' psyche who avoided any direct confrontation with the RR troops. Secondly, due to the pro-active nature of operations conducted by well trained troops, the militants lost cadres, arms/ equipment which were quite a setback to them. It is only the RR troops who have been able to maintain such pressure. The motivation to perform well has its impetus in various factors such a sense of pride to get selected in a special force with separate identity, dress and organisation. Each individual therefore looks forward to prove his mettle in the operational field.

Weaknesses

Not withstanding the commendable operational performance of the RR, some senior military leaders have asked for a re-look at the RR. According to them the RR has served its purpose and now it is time to restructure it as part of the Internal Security Force. The reasons attributed by him and many others relate to the negative trends that have developed over the years and adversely impacted the overall ethos and discipline on the Army.

- **Ethos of RR**. The RR is nothing but the Army with a different designation. Its training and equipment are the same as that of the Army. The men too are from the Army with the same uniform but a different badge and insignia. Only its modus operandi and tasking is similar to the CPO forces. Therefore, the common man's perception of the Army has been clubbed with the CPO forces. More importantly the employment of the Army as an anti militancy force has come to be a reality. Secondly, the RR personnel because of the nature of duties/ tasks they perform, imbibe the ethos of the Police which is characterised by loose discipline. This ethos is carried with them when they revert to their parent units. Thus, over a period of time, this ethos of loose discipline has permeated into most of the units of the Army.

- **RR as PMF**. The RR is a PMF only in name. Its training and equipment are akin to the Army. The RR is under the Army and funded by it. So, how can it be designated as a PMF? Whom are we trying to fool?

- **Command and Control vis a vis the CPO Forces**. One unforeseen consequence of RR being a PMF is that it has led to command and control problems with the CPO forces. Like the BSF and CRPF, the latter started arguing that RR is a junior organisation. Therefore, they cannot be placed under their command even for operational purposes even though they are commanded by regular Army Officers.

- **Cohesion and Integration**. Not withstanding the fact that the RR battalions are now affiliated / part of one of the regiments, their manpower is still being provided by different battalions of that regiment. Secondly, the other two companies continue to be part of Armoured Corps/ Artillery regiments. The personnel of Engineers, Signals and ASC being still part of the RR battalion further compounds the problem. All this mix of personnel, poses the problem of RR battalions being fully cohesive and integrated unit. They also fail to develop the necessary espirit-de-corps.

• **Suicides and Fratricides.** The RR battalions were not only the initiators of the spate of suicides and fratricides that followed in the rest of the Army units but also the major contributor. This is being attributed to the lack of cohesion and the ethos of the CPO forces having been imbibed by the RR units. This is because the RR has followed the Non Combat Model, thereby its individuals are rotated from one unit to the other. Thus there is no regimentation. The reorganisation of this force has not been accompanied by the process of socialisation so vital a factor for developing the Espirit de Corps. This bond between the soldiers fighting the militants/ terrorists on the front line is non existent. This is a failure to create a social group in the units and sub units, based on common values. The simple process of the four Cs (Care, Concern and Commitment leading to the Fourth C, Camaraderie) is missing. For inadvertant liberation such regimentation and bondage is anathema as it imprisons the soul which in the line of fire is the most humanising influence in the face of constant prospects of death.

• **Turbulence due to annual turn over.** 55 percent personnel get tuned over every year as part of the rotation. This coupled with the frequent change of leadership causes severe turbulence in the RR units.

What is the solution?

To overcome the above problems, suggestions have been made to restructure the RR force as part of the Internal Security Force. This issue is not pertaining to the research and thus not being dwelt upon any further. From the point of this research, it is important to find suitable answers to overcome the weaknesses in the RR force so that there is a reduction in the cases of Suicides and Fratricides.

Conclusion

The RR is a potent force and an answer to relieve the Army from getting involved in CI Operations. It requires further refinement to prove its full utility.

Section 3 – Women in the Army

"Changes in mindset and attitudes are required to fully accommodate women in the Army"

Introduction

The women have been in the Army for almost two decades and proved to be an asset. Their induction has overcome to some extent the acute shortage of officers in the Army. However, there have been problems which need to be highlighted and over come. The suicide by five women officers in one year alone has necessitated examination of some of the issues that affect this class of officers.

Historical Background

After the signing of the Armistice Agreement in 1919, the process of discharging the women began and by 1922, all women had been discharged.

During WW II, after the attack on the Pearl Harbour, the women were again recruited as part of the Women Auxiliary Corps (WAAC) which later became WAC. Later, women were accepted for Voluntary Emergency Services (WAVES), the Women Reserve in the Marine

Corps and the Coast Guard Women Reserve (SPAR) in 1943. During WW II, a total of 1,50,000 women served in various capacities like clerks, health care specialists, airplane mechanics, gunnery and instrument flying instructors , ATC, airplane ferry flight pilots flying all types of aircraft. About 800 women Air Force Service Pilots (WASPS) flew all types of aircraft including combat planes from one base to another. They were however not given the full military status as Aviation pilots. (Yet one woman pilot was killed in action due to enemy artillery fire).

Between WW II and Vietnam War, the proportion of enlisted women again fell down to 1.1-1.5% from the 2% quota that existed until 1967. At that time, women were enrolled to overcome the shortage of manpower required for the Vietnam crises and the quota was lifted. Women enlistment was not realised until 1972 when there was substantial increase in the enlistment of the women to offset the shortfall in men's enlistment with the ending of conscription. With the passage of the Equal Rights Amendment Bill, the military sought to enhance its image as a model of equal opportunities by enlisting more women and assigning them to all occupational jobs, except those relating to Combat Arms. This substantially increased the proportion of women to 11% of the total enlisted force. By 1974, the Navy had 7%, Air Force 15%, Army 22% and 14% Marine Corps. Since then, women in the Army are a reality.

From the above historical perspective, it will be seen that women have been enlisted in the armed forces for mainly two reasons:-

- Make up the deficiency of men or release them for combat duties.

- Give equal opportunities to the women.

Increased participation of women has been a good experience. Their increased participation with higher mean levels of education, general aptitude, age and socio-economic status has enhanced the over all quality of the all volunteer force. It can be concluded that the role of women has been beneficial and should be expanded significantly.

The Indian Experience [40]

The Army has been confronted with a shortage of approx 13,000 officers. To tide over this problem, the idea of recruiting women in the Army was mooted by Gen S F Rodrigues, the then Chief of the Army Staff. Consequently in 1993 the doors were opened for recruitment of women in the erstwhile all male bastion. The eligibility for women officers is:-

- Age   -   19 to 25 years.

- Height   -   142 cms.

- Weight   -   36 Kgs.

- Medically fit.

The qualification was dependent upon the type of entry.

- Graduate and Post Graduate

- Technical Graduate and Post Graduate.

- Especially Skilled.

The entry is open to ASC, AOC, EME, ENGRS, SIGS, AD Regt, Legal, Intelligence and Army Education Corps. This was besides the Army Medical Corps in which women were already allowed as doctors and Nursing Officers.

Two years anti date is granted for technical graduates on commission. All are granted the rank of Lieutenant on commission.

---

[40] Women in the Army by Vipin Agnihotri-httr:ezinearticles.com/?.women/ in the Indian Army and Aid=253178&opt and an Article,An Officer and a woman by Rajat Pandit- *Times of India,03 Jul 07.*

They have to undergo six months of training at the Officers' Training Academy at Chennai as opposed to nine months for male officers (This has been made equal to that of male cadets) On completion of training, they are granted short service commission for five years extendable to 14 years. (They have now be offered Permanent Commission). Their starting salary is in the scale of Rs 8200-10050, total being Rs 16,000(*This has been enhanced after the 6th CPC*). Any officer seeking release earlier than 5 years had to refund the cost of training at Rs 7500 per month.

The response has been good. Almost 5000 women officers, both Graduates and Post Graduates apply every year. Barely, 1/10th are successful through the selection process.

When asked as to what motivated them to join the Army, their response indicated that these women had different motivations:

- Majority of them were seeking a service career.

- Some had a sports oriented outlook and thought that the Army would provide the opportunities to further their interest. Others were looking for adventure like mountaineering.

- All of them looked forward to tenure abroad as part of the UN contingent.

- The widows felt that they needed to carry on from where their husbands left.

- Some women were looking for financial benefits like Provident Fund and savings.

Whatever be their motivations, the women in the Army are quite motivated, dedicated and determined. To-day there are 1,468 (excluding MNS) women officers in the armed forces. The Army's share is about 1000 out of a total of around 40,000 officers.

Having joined the Army, how do they feel? Have they been accepted in the system or are there any prejudices? The answer was reflected in the statement by the top leaders in the Army. Some retired Generals also expressed their views through the articles.

The reality is that there is bias against the women officers. The bias was reflected in a statement made by none other than Lt Gen S Pattabhiraman ex vice chief of the Army Staff who said "They could do without them". This statement drew a lot of criticism and reaction not only from the National Commission for Women but parliamentarians like Sushma Swaraj of BJP who asked the government to tell the General that it could do without him. Though the clarification and apology was rendered by the General, the statement did reflect the attitudes of some of the Army officers. A similar view was expressed by Lt Gen R S Kadyan, retired as Deputy Chief of the Army Staff in his article in the Tribune "Women Officers in the Army" .He argued that women officers required special facilities of living to maintain their privacy and it was not possible to provide the same. He also remarked about their physical inability to undergo hardship.

When some of the women officers were asked the question about discrimination, they replied, "Discrimination was not the issue. It is the gender bias which is reflected in the attitudes of seniors, colleagues and the men". According to them their induction is quite an uneven experience. While they complain of condescending attitudes from seniors and colleagues, of cultural and practical problems of having women in the Army, their capabilities are questioned or under doubt even when they out perform. Their greatest challenge is to prove that they are no less in performing any task including that of physical strength and endurance. There are also the problems of choosing a second career in mid life with no pension being granted. Apart from questioning of professional acumen, there are cases of sexual harassment (10 cases since 2004). Male officers crib about women officers being "Molly Coddled" are soft duties being assigned to them. Seniors seem to be too scared of harassment. They are thus over protected. Women officers tend to by pass, seek preferential treatment in postings, duties and leave.

Women join without being fully aware of the hardships in the Army. The threshold of some women officers is less and they breakdown fast under pressure. However, women officers dismiss all such notions. They accept the fact that there are adjustment problems in the early stage but most of them make the transition into service ethos. They are highly disciplined and cultured properly.

From the above, it is evident that some of the Army Officers have not yet accommodated the women into the fold of the army. They still have a mind set which is prejudiced against them and their attitudes reflect the gender bias. This is particularly applicable to the men. Unless the mindset and the attitudes change, women cannot be fully accepted and accommodated in the Army.

What about their aspirations? Have they been fully met? The answer is simply NO. The women aspire to join the Combat Arms and retire as permanent commissioned officers with full pension. Therefore, while majority of the women officers have adjusted and are quite happy, some of them still have not been able to reconcile their aspirations. And there are still some who have not adjusted to the hardships of the Army career.

The other issue that has agitated the minds of many is the suicide by women officers. There have been five cases so far. Some of the causes attributed to these cases are:-

- Harassment by senior officers.

- Unhappy with the duties assigned to them. These were not in keeping with the qualifications they had. Dissatisfaction with the Army and inability to obtain release and compensation.

Conclusion

In conclusion it can be said that there are still mutual biases. But the induction of women officers in the army is a reality. Therefore, for the betterment of the Army, the sooner the male officers and particularly the men change their mindset and shed their gender bias the better. On the other hand, the women officers too must learn to adjust and be prepared to go through the hardships of army life. Lastly, their minimum aspirations for a permanent commission and pension on retirement, need to be looked into.

CHAPTER 10-SYSTEMIC ISSUES

Having processed various studies on the subject of psychological effects of LIC and interacted at various levels with both –the serving and retired army personnel, it came to light that there were some fundamental issues at the macro level (Systemic Issues) which were not addressed but were greatly responsible for the increased pressure on the troops, leading to combat stress.

These issues are being discussed in subsequent paragraphs.

Absence of National Doctrine and Strategy of LIC

A questionnaire was issued to the higher hierarchy of the army (Divisional, Corps and Army commanders), regarding some larger issues concerning management of LIC. The aim was to ascertain whether there was any national strategy/ military doctrine laid down at the National/ Army headquarters level or not, so as to clearly initiate a well thought out policy and achieve cooperation/ coordination of all the agencies to resolve the problem of LIC in a particular region or state.

The reply was revealing. There has never been any directive issued by the Ministry of Defence to the Army Headquarters. Similarly, the Army headquarters too has failed to lay down any directive and therefore spell out a clear military strategy for the Army Commands. Ex Army commander- Lt Gen Rustom Nanavaty says- The non existence of a comprehensive strategy and the absence of an effective institutional framework for the government resources and efforts are the two main obstacles to the efficient management of the conflict. [41]

In the absence of any national policy or Army Headquarters directive, it is only a few Army commanders, both in the NE and J&K, who laid down their own guidelines for the lower formations. In most of the cases, the commanders at different levels (Army Commander downwards), were conducting operations based on their own perceptions and previous experience. Since, most of them were exposed to LIC only at the battalion and company levels, these commanders as divisional/ corps commanders were conducting LIC operations mainly at the tactical levels. In sum, a few Army/ Corps/ Divisional commanders were fighting the battles at the tactical levels (Company/ battalion commanders), thus laying emphasis mainly on the numbers game.

It is therefore no surprise to see that most of the commanders were conducting LIC in J&K, using the same template of the methods employed in the NE or OP PAWAN. Axiomatically, strategic issues like strategic intelligence, management of environment and the media and coordination of all security forces is over shadowed by the tactical methods employed.

Assessment of units and individual's performance is also evaluated based mainly on the results produced criteria.

Undoubtedly, there has to be some method of evaluation and there cannot be any better than the achievements in tangible terms. Therefore, at the unit level and tactical level- optimum utilisation of full potential of the nation, it ultimately boils down to this only. This argument holds good and has merit provided the element of the

---

[41]   Extracts of Seminar on J&K and Contours of Future Strategy held in CLAWS on 02 Jun 07

local population is not ignored. After all, the Army doctrine is based on winning the hearts and minds of the local population. Therefore, this cardinal principle cannot be ignored or overlooked while achieving the results.

Absence of Institutional Framework for Management of LIC

The second issue relates to the absence of an institutional framework at the national level. As per the constitution, law and order is a state subject. Accordingly, the state police should handle the LIC. But regrettably, despite so much emphasis to better equip and train this force, it remains incapable of handling the situation. Under such extraordinary conditions where the situation goes beyond the level of just a law and order situation, it becomes incumbent on the Union/ Central government, to release its resources of CPO forces and the Army. The CPO forces also being incapable of containing the LIC, the burden of managing the LIC, falls on the Army. Thus, there is a dichotomy in the system. The state which is responsible for law and order does not have the resources to fulfill this responsibility and the army and the CPO forces, which become the leading agencies in containing the situation, are not accountable to the state but to the centre. Ipso facto, it is the Central government that becomes accountable without being directly responsible. Such a paradox brings about direct and increased pressure on the army. It has to do so besides its primary role of safe guarding the national borders against external aggression.

The problem gets further complicated when the state police which is closest to the ground realities does not operate under the lead agency, the Army. Similar is the case with the CPO forces. Except for the operational control, the Army does not retain the administrative control over the CPO forces. The CPO forces therefore are not as responsive to the orders from the Army as they should be. Thus, there is losing control of the Army on the state police and the CPO forces, resulting in lack of applying the full potential. The Army continues to bear the main burden with increased pressure on its officers and men. This applies equally to the full use of all other agencies/ organisations in a coordinated manner. This highlights the need for an institutional framework at the national level for better management of the LIC. From the above, it is evident that the army alone bears the main responsibility of containing the situation. In its exuberance and traditions to deliver, the men come under tremendous pressure. Whether this pressure is self generated by ambitious unit commanders or it permeates from top is an academic discussion. The fact that this pressure gets multiplied with prolonged deployment of army in LIC operations is the reality. The Army under increased pressure, therefore, is under stress.

Shortage of Officers

Against an authorisation of 46,615 officers, there is a shortage of 11,235 Officers. This shortage is mostly at the junior level (Majors and below). According to the statement made by the Chief of the Army staff on the occasion of the Army Day on 15 Jan 2008; only 190 candidates out of a total vacancy of 300, reported for training at the National Defence Academy and only 86 joined the Indian Military Academy against 250 vacancies. This is mainly due to the Army being the least preferred option as a career. The problem is further compounded by officers seeking discharge to find better job opportunities in the Civil (Multinational companies and Corporate) Sector.

In 2006, 811 officers applied for release/discharge. This shortage has remained with the Army for over the last decade. It not only adversely affects the conduct of CI operations but also the Man Management in the units with serious consequences as experienced these days. (Refer to an Article, Officers Leaving Forces Despite Pay Hike Promise. rajatpandit@timesgroup.com)

Shortage of Family /separated Family Accommodation

The reality today is that jawans and their spouses wish to stay together as long as they can, particularly when the unit comes to a peace station. This has been necessitated due to most of the families being nuclear. The increase to even 35% scale is not adequate to meet the total requirement. The CILQ (Compensation in lieu of Quarters) is not adequate to get reasonable accommodation outside the unit. This deprives the jawans from staying with their families for the full peace tenure. A jawan barely gets to stay for one and a half years out of the three years tenure. Even in stations like Delhi, an officer gets permanent accommodation after two/three shiftings in temporary accommodation and barely gets a maximum of two years to stay in the permanent accommodation before he gets posted out. If this is the state of accommodation of officers posted to the Army Headquarters, then one can imagine the plight of JCOs and ORs in other stations.

The situation with regards to separated family accommodation is still worse. This is the time when the families want to be in a secure place and be able to give better education to their children. But the non availability of adequate separated family accommodation compels the families, especially JCOs and ORs to either stay in their native place or unhealthy surroundings with attendant problems. This puts a lot of pressure on the men operating in forward areas leading to stress and suicides.

Empathy of the Government Officials

There is total empathy on the part of the government officials in addressing the problems/ grievances of the soldiers. Despite a personal talk by the Union Defence Minister while addressing the Chief Secretaries of the states on this issue, the situation has not improved except probably in the state of Haryana where the Chief Minister has directed that all cases of the soldiers and Ex-servicemen will be disposed of on priority. In some states not even the letters by the commanding officers are acknowledged. This leaves the soldiers frustrated and puts pressure on their minds, leading to suicides.

State of CPO and State Police

The CPO forces and the State Police are neither equipped nor trained sufficiently to deal with the militants and the terrorists. These forces lack latest weaponry, equipment especially communication facilities and vehicles for mobility. They also lack sufficient training in dealing with militancy. In such a situation the government has no choice but to put the whole responsibility of containing the militancy on the Army. This has led to increased and protracted employment of the Army resulting in the present situation of increasing suicides and fratricides.

Use of the Military as the final option to deal with Militancy

The cause of any militancy or insurgency is political, economic or social. The emphasis is therefore more on resolving the issues through political, diplomatic and other means. The Army can only be used as a sub median rather than a final option to resolve such complex issues. The Army can at best contain the situation and create favourable conditions for talks. But invariably, once the situation is handed over, complacency sets in into the political leadership and the issue is relegated. Even when the situation is favourable, the government neither tries to function nor makes serious attempts to resolve the issue through other means. The best the government does is to appoint a bureaucrat to progress talks with the local people. Such a situation of political inaction and stalemate frustrates the soldiers with grave adverse consequences.

Conclusion

The Army is a national asset. Its problems therefore are not its own. While it deals with the Non- Systemic issues, the Nation, both at the centre and at the state level must address the systemic issues.

CHAPTER 11-MANAGEMENT OF STRESS

"Our whole life is taken up with anxiety for personal security with preparations for living, so that we never really live at all"
- **Leo Tolstoy**

Introduction

As mentioned earlier, Stress is not pressure from outside situations called Stressors but responding to such situations/stressors. Such a response results in emotions like Anger, Anxiety and Frustration. Stress begins with being anxious because of workload: this causes Anxiety which leads to tension. This tension triggers nervous impulses that cause physical changes in the body. And when Tension reaches a point that affects the body, people are under Stress.

While stress cannot be totally eliminated it can be managed and minimised. The following should be controlled to manage Stress:- [42]

- Anger.

- Anxiety.

- Tension and

- Time.

Management of Anger. Common emotions can be dangerous. Anger is a common emotion, can lead to destruction of property, disruption

---

[42] Hand Book on Management of Stress issued to the Army.

of relationship and even death sometimes. There are three steps to control Anger:-

- Relaxation.

- Positive self talk.

- Remind yourself-What is it that I cannot do? ; Take Anger as a signal- I am slightly angry- Time to relax now, Keep the goal in sight not Anger- I can do it and I have done it before.

Management of Anxiety. What we say to ourselves changes what we see, hear or feel (Relaxation Technique).

- Remind yourself. You are in a situation where you are likely to become anxious. Tell yourself- What is it that I need to do first?

- Take anxiety as a signal. Tell yourself something positive. I am feeling a little tense.

- Keep the goal in sight, not fear. I need to do it. I can do it. I have done it before.

- Pursue. My body is tense, my mind is anxious but I can overcome. Let me do one thing at a time. I am in control. I need to do a little more.

- Reward yourself. After some reduction in anxiety say-I am confident now. I can do better.

Management of Tension.

- Relaxation Training- Relax the mind by relaxing the body.

- Imagery Training- By experiencing through all five senses an imaginary pleasant scene.

Managing Time.

- Make a List-

  (i) What must be done?

  (ii) What should be done?

  (iii) What would you like to do?

- Cut out time wasting activities.

- Learn to drop unimportant activities.

- Say NO or Delegate.

Other Activities.

- Healthy Eating Habits. Avoid too much caffeine and salt.

- Moderation in consumption of alcohol, smoking and NO use of Drugs.

- Exercise, Good sleep and Relaxation (Difficult to cope with stress when tired. Sleep is a good stress reducer. One wakes up refreshed after a good night's sleep and has plenty of day time energy. Leisure / Recreation provides outlet for relief and gives break from stress).

Managing Stress

'The mind uncontrolled and unguided will drag us down, renders us, kill us and the mind controlled and guided will save us and free us" - Swami Vivekanand

_____

Also refer to Articles ;Fighting Stress- Sat, Extra, *The Tribune ,06 Jun 2007* and Reducing the Work Stress Among Personnel – Press Information Bureau, Government of India, 07 Mar 2007.

Organisational Measures.

- Motivation.

- Increased communication.

- Improve quality of life.

- Improve working environment.

- Introduce formal education in Stress Management.

- Review Qualitative Requirements (QR) for various appointments.

- Educate Spouses.

- Organisational Support.

- Optimize Self Esteem.

Stress Management in LIC/HAA/CI Environment.

- Keep Troops informed and in picture.

- Maintain the Weapons and Equipment (At par with the Militants).

- Improve Hygiene Conditions.

- Proper Leave Management.

- Correct Leadership.

Tips for minimising Stress at Individual Level.

- Don't create stress for self.

- Ensure Financial Security.

- Be Yourself.

- Never Compare.

- Improve your ability to communicate.

- Change Gears.

- Cut back on Excessive hours.

- Pamper Yourself.

- Practice relaxation techniques.

- Forgive and move on.

- Learn to accept criticism

- Manage your time better.

- Seek help when under stress.

Role of a Leader as Stressor

A leader can be a Stress creator if he:-

- Sets unattainable Goals.

- Violates the chain of command.

- Unable to provide hygiene facilities

Leader as Stress Reliever. A leader can also be a stress reliever if he:-

- Creates a healthy working environment.

- Wins the trust of his Command.

- Is accessible.

- Rewards good work promptly.

- Maintains equanimity even during reverses.

- Looks after hygiene factor.

Golden Principles.

- Laugh rather than cry.

- Praise rather than decry.

- Create rather than destroy.

- Heal rather than wound.

- Give rather than grab.

- Pray rather than curse.

- Act rather than procrastinate.

- Grow rather than stagnate.

- Love rather than hate.

- Live rather than exist.

Conclusion

Before conclusion, it will be of relevance to narrate an anecdote that was narrated by an NRI during her visit to India and deals with stress management. "A lecturer when explaining stress management to an audience, raised a glass of water and asked:-

How heavy is this glass of water?

Answers called out ranged from 20g to 500g.

The lecturer replied, "The absolute weight does not matter."

It depends on how long you try and hold it.

If I hold it for a minute, that is not a problem.

If I hold it for an hour, I'll have ache in my right arm.

If I hold it for a day, you'll have to call for an ambulance.

In each case, it's the same weight, but the longer I hold it, the heavier it becomes.

He continued, and that's the way it is with stress management.

If we carry our burden all the time, sooner or later, as the burden becomes increasingly heavy, we won't be able to carry it.

"As with the glass of water, you have to put it down for a while and rest before holding it again.

When we are refreshed, we can carry on with the burden"

So, before you return home tonight, put the burden of work down.

Don't carry it home.

You can pick it up tomorrow.

Whatever burdens you are carrying now, let them down for a moment if you can.

So, my friend put down anything that may be a burden to you right now.

Don't pick it up again until after you have rested a while.

CHAPTER 12 - MAIN OBSERVATIONS/ COMMENTS BY THE AUTHOR

General

The Army is suffering from the malady of Stress, Suicides and Fratricides for over the last decade. Several studies have been carried out within the Army and outside it, by various sociologists and psychologists and remedial measures recommended. The Army and the MOD have initiated and instituted several measures to prevent or minimise such cases.

The Army has initiated measures in several areas to address these issues. Some of these are:-

- There is greater emphasis on Leadership, Leader-Led relationship, Man Management, and Motivation and Morale. Leaders at all levels have been sensitised.

- Adequate number of JCOs/NCOs has been trained as counsellors. Committees to review and analyse Stress related incidents have been constituted and tasked to suggest additional measures to be taken. Institutionalised training is being imparted to officers, JCOs and NCOs at various courses of instruction.

- Yoga and other exercises have been introduced in the daily routine to manage Stress.

Similarly, the MOD has introduced several welfare measures to alleviate the sufferings of the soldiers.

From the above, it is apparent that the Army is doing all it can possibly do to address these issues. It has also its own system of checks and balances to prevent occurrence of such cases. There is also no denying the fact that there is no parallel in the world which is comparable to the achievements of the Indian Army. While other major armies like the US Army in Vietnam and the Soviet Army in Afghanistan failed, the Indian Army has an excellent record of having succeeded in containing the insurgency for over five decades. It is also a fact that the Indian Officer is known for his professionalism and unique quality of leading from the front. He has acquitted himself with honours and made the supreme sacrifice of his life as and when the situation has demanded. To that extent, he has done the country proud. Similarly, the Indian Soldier is known for his excellence in braving the hardships of harsh terrain and climatic conditions and withstood the privacy as well as deprivations in life without a grudge. It is also a matter of pride that despite the adversities and hard life, the Morale of the soldiers is high and the unit/subunit cohesiveness is fully intact.

Notwithstanding the above, there is another reality-the cases of mental disorder, Suicides and Fratricides are still taking place. The worry is not why such cases are happening? They have been taking place even earlier and will continue to take place. This is acceptable. The issue is really why they are taking place at this large scale? The worst is that despite so many remedial measures by the Army and the MOD, there is no perceptible decline in such cases.

This is a reality and we cannot ignore it. The soldiers are giving a clear signal that there is still some malaise somewhere that needs to be addressed.

May be the Army and the MOD do not deserve the type of indictment as has been given by the JPC in its report of 2008-09. But it does raise some serious questions.

- Has all been done that was to be done?

- Have the measures recommended been implemented with seriousness?

- Is there still any malaise that we have to examine to address?

The soldier is also giving a clear signal through committing Suicide or killing a member of the Fraternity. The Army can no longer ignore these signals. This is an important fact; it not only recognizes the problem but also its gravity.

Main Observations

Having given an understanding of various maladies, the factual data and their analyses, there are some observations/ comments by the author. These are contained in subsequent paragraphs. There are some factors which are accentuating and those that contribute directly to stress. These must be clearly understood. Some of these are highlighted in subsequent paragraphs.

Impact of Environment

Environment is one of the major factors that play an important part in inducing stress. Its various dimensions are as follows.

Impact of TV/DVD. Troops spend a lot of time in watching TV and movies. This influences their minds and raises their aspirations and expectations. They start comparing themselves with their compatriots in the society and feel left out. They feel that they are not being fully compensated for the jobs they do and the risks they take in ensuring the territorial integrity of the country. Similarly, the possession of the mobile phone is a double edged weapon. Their use needs to be controlled.

Impact of the Society

- **Officers.** Not withstanding the aberrations that have taken place in the character qualities of some of the officers, by and large, they still care for their men. The real contact level being at the junior level, they do not find adequate time for their administration and personal/domestic problems because of the acute shortage at this level.

- **Soldiers.** They are of two categories- those hailing from the traditional area (Rural) and others that are from the Non Traditional area-(urban). Since majority constitutes those from the rural areas, the Indian soldier is generally simple, frugal and can brave a lot of hardship. Their demands are few. However, they are intolerant of insult and injustice. On the whole, there are an increasing number of Nuclear families raising the levels of burden of responsibility on the uniformed men. The wives of a lot of soldiers are well educated with high expectations of an Army life. Possession of Mobile especially by those who are separated, complicates the issue. There is no doubt that because of higher aspirations, everyone wants to achieve the highest rank in his/her category. This has manifested itself in all ranks, becoming conscious of their ACR and is ready to compromise to get a good one. Lastly, since they are a product of the society, they imbibe the negative traits of being intolerant, impatient and materialistic- this leads some of them into debt.

Environment in the Army. The Army has its own routine, is known for keeping the area neat and tidy. It has a unique culture of a close bondage with everybody leaving like a family with the commanding officer being the head of the family. Everybody speaks Hindustani and is an example in national integration. There is no distinction of religion or caste. It follows the chain of command for communication, upwards/downwards. The environment in Peace differs from that of the Field Area (LC, CI, HAA or Snowbound area). The environment of LIC is described briefly as under:-

- **Environment in LIC**. The environment in LIC and its nature of operations are very challenging-Fighting with constraints, Hostile Population, Adverse Media, and Pressure to perform-Harsh climatic and Terrain conditions coupled with Ad hoc living conditions-Separation from the family- Large Areas to be dominated and frequent moves and redeployment, constant vigil against uncertainty of attack- inadequate time for Rest/Sleep and irregular meal timings. Add to these, the feeling of isolation and the fear of losing a limb or life. All these factors put a great psychological weight on the minds of the soldiers, resulting in the loss of mental balance. The operational climate and the environmental conditions of LIC coupled with the problems from his family side put a great strain on the soldiers.

- **Environment in Peace**. The environment in Peace Area is no better. There are far too many training, sports and social events. In addition, the soldiers are required to do so many non- professional jobs. The soldiers are so busy that they do not get adequate time to spend with their families and take care of their problems.

From the above, it is apparent that whether in field or peace, the Army is so busy that the soldiers are under a lot of pressure. Therefore, the first step that needs to be taken is to reduce the pressure.

Stress

Stress is the mother of all other maladies like suicides, fratricides, Alcoholism and mental disorders. There is no doubt that the life in the Army is tough. Therefore, whether in peace or field, stress is inherent. But why is stress inherent? What are the causative factors of stress?

Causative Factors of Stress. From the evaluation of various studies, analyses of the data and interaction with patients with mental disorders, it is apparent that there are different causative factors. In the CI environment, it is mainly a by product of three groups of stressors-

- **Operational/Occupational Stressors**. (Anger at Fighting with constraints - Zero Error Syndrome-Fear of ever present danger- Fear of Uncertainty- Fear of losing a limb/loss of life)

- **Environmental Stressors**. (Pressure of work-Paltry pay & allowances- Unhappy with food and Living conditions). Though the Army should not down play the importance of providing good food, good living conditions and better pay and allowances, the main issue is lack of adequate rest / sleep and irregular meal timings. Whether the Army accepts it or not, the reality is that an individual gets not more than 4-5 hours of sleep and maximum 2-3 hours for personal administration. The tempo of operations is so fast and the pressure to perform so high that the individual has irregular meal timings. *There is therefore a strong linkage between the Operational and the Environmental Stressors.*

- **Personal /Domestic Stressors**. (Marital discord-Jilted in love affair- Financial hardship-Bypassed in promotion and so on). With the break up of the joint family system, there is an increasing number of nuclear families. This has imposed direct responsibility on the individuals. The matter gets aggravated when the individual is posted to a field area, especially to the LIC Environment. There is a separation from the family and the children have to manage in rural schools. In today's context, wives of some PBORs are better educated and their aspirations are high. Possession of mobiles by some of them, aggravates the problem of the individuals. While there is constant fear of losing his life or limb, there are pressures from the family to resolve their problems. The individual can proceed on leave on some occasions but there is a limit to it.

The evaluation of stressors indicates that the Operational Stressor is No1, Environmental stressor is No. 2 and Domestic stressor is No 3. But there is such a strong linkage between the Operational and Environmental stressors. Therefore calling them as No. 1 or No. 2 would be purely a matter of semantics.

Suicides

It is good to compare the average number of suicides in the Army with those of the National Average as also with the average of other armies in the world like the US, UK, France and so on. But it would be wrong to draw comfort from the fact that our average is less because neither can there be a comparison between a civilian and a soldier nor can there be a comparison between the Indian soldier with those of other Armies as the socio-economic conditions, the cultural background and the threshold of tolerance levels of the Indian soldier are quite different.

Peace vis-a-vis Field. There is a belief that there are more number of suicides in peace than in the field. While this is true but on Pro-rata basis, the average number of suicides in peace is less than those in the field-(11:13/100,000).

J&K vis-à-vis NE Region. Similarly, there is a belief that there are more cases of suicide in J&K than in the NE Region. This again is not true as on Pro-rata basis, the number of suicides in J&K and NE Region are generally the same.

Suicides in units/castes. There is an erroneous impression that soldiers belonging to a particular caste or regiment are more prone to suicides than others. This is not true. The overall figures only indicate that the troops from a particular region/caste or community are more sensitive and therefore vulnerable to committing suicides. It is for the leadership to- understand how such troops need to be handled. Similar argument holds good for units of RVC, AOC and DSC, who have highest number of suicide cases on Pro-rata basis.

Causative Factors of Suicides. The real cause of suicide cannot be ascertained in 70-80 percent cases. However from the trends and some evidence that may be available, the following major causes are attributable to suicides:-

- **Domestic Problems (Familial/Personal)**. This includes cases of marital discord, jilted in love, parental conflict, and other personal problems. These constitute the maximum cases.

- **Mental Disorder (Depression)**. There are several cases that have the genes, are prone to depression and have suicidal tendencies. If there is a psychological screening at the recruitment stage, then they could be detected and rejected. Such cases cannot withstand pressure /stress and tend to go into depression.

- **Medical Problem.** When an individual suffers from a medical problem like dysfunction or HIV and when such a problem is incurable, the individual may commit suicide.

- **Debt.** A large number of individuals get into heavy debt due to various reasons. When such individuals realise that they cannot pay such heavy debts, they put an end to their life.

- **Non Grant of Leave**. A soldier is emotionally attached to his family. When there is a problem in the family, he likes to be beside them to help and would desire to go on leave. If leave is not granted at such an occasion for whatever be the reasons and he is not able to go, he gets a feeling that he is of no use to his family. In such a situation, he considers himself as useless since he cannot be of help to his family when needed; he considers his life not worth living and commits suicide.

- **Self Respect**. There are cases like sodomy when the individual either as a victim or perpetrator of the act feels shameful and cannot show his face to his colleagues. In such a situation, the individual puts an end to his life.

- **Other causes**. These include cases like failure in Promotion examination, being superseded, or not getting selected as an officer. Any type of failure to get something an individual aspires for could lead to suicide.

A case of suicide or an attempt to commit suicide, could be prevented by keeping a close watch on the individuals But whether it is possible to identify such a person and prevent him/her from committing or attempting suicide is a debatable issue. Another issue that needs to be debated is whether such a person should be retained in service or be medically boarded out ? The main point at issue is that the Army cannot afford to lose a trained soldier because of his or her making an attempt to commit suicide.

Lastly, a case of Self Inflicted Injury should be treated as a case of an attempt to commit suicide and must be reported.

Fratricides

Fratricides is a serious problem and are a consequence of an individual's intolerant/impatient nature. This is a reflection of the society where cases of road rage are not uncommon.

Inter Personal relations play a vital role and the Army has a culture of its own where there is a strong bondage amongst various ranks like a family. Some how this fabric gets disturbed-where such bondage does not exist due to adverse Inter Personal Relations. In such a relationship, even a slight provocation can invite a strong reaction especially when every one is under stress. In such cases the reactions become more "Emotional" than "Rational". The possession of weapons and ammunition facilitates Fratricides. This is why there are more cases of Fratricides in the Field – both in J&K

and the NE Region. And such cases are more in J&K than in the NE Region. Similarly, there are more cases of Fratricides in the RR Battalions in J&K and AR Battalions in the NE Region than the Infantry Battalions on Pro rata basis. As in the case of suicides, troops of the Northern Region and accordingly some communities are more sensitive and therefore get easily provoked. Maximum cases of Fratricides are against NCO's/OR's, then JCO's followed by Officers. It only shows that the PBOR's still respect their Officers.

Cases of Negligent Discharge of Weapons (NDW) could be an attempt at Fratricide and should not be dismissed as mere accidental firing.

There are mainly five types of provocations that result in Fratricides:-

- **Humiliation.** (It could be through Insult, using abusive/ Intemperate language or public punishment)

- **Being insensitive to the sensitivities of an individual.** (Hurting somebody's religious feelings, beliefs or taunt at somebody's weakness)

- **Injustice.** (Injustice in grant of promotion or unjust distribution of duties)

- **Non grant of leave in an emergency.**

- **A mere argument spinning into a fight.**

Other Maladies

There are two more maladies that are generally not discussed or talked about. These are Alcoholism and Neurosis.

Alcoholism. The Army has exercised a lot of control on the consumption of Alcohol after the issue of AO3&11 /2001. However there is a problem in peace stations where exercising control on the consumption of Alcohol is not possible due to certain unavoidable reasons. (Consumption of Alcohol in family quarters especially when entertaining a guest). This has led to more alcoholic cases in peace stations than in field area.

Neurosis. A study on Neurosis suggests that the families of service personnel are as vulnerable to Psychiatric disorders as the service men. That is why there have been cases of suicides, even in the families. This calls for counseling of families in peace stations and separated family accommodation.

Impact

The Army would realise the gravity of the problem when it seriously views the human cost of cases due to mental disorders, suicides and fratricides, There are more cases of Psycho/Neuro and Alcoholic cases than those killed in encounters with the militants. Similarly, the number of lives lost through suicides and fratricides or rendered mentally unfit due to mental disorders is twice the number killed in encounters.

CHAPTER 13 -REFLECTING ON THE REMEDIAL MEASURES

General

The Army is a national asset. It is therefore not the sole responsibility of the Army alone to find remedies to the malady of Suicides, Fratricides and Stress. The nation too has to also share the responsibility to take corrective actions by acting in unison. There are issues which are within the realm of the Army and others that are beyond it. It will therefore be prudent that issues, both systemic and non-systemic be dealt with at respective levels.

The Ministry of Defence (MOD) has already announced the following measures: -

- Enhancement in insurgency allowances.

- Grant of railway warrant twice a year.

- Appointment of psychologist counselors and augmentation of psychiatric centres in J&K and NE Region.

The above measures are steps in the right direction. However, there is a serious indictment by the Committee on Defence (2008-2009). In its 31st Report on 'Stress Management in the Armed Forces", the Standing Committee has concluded-[42]

---

[42]   31st report of the Standing Committee of the Lok Sabha on " Stress Management in the Armed Forces."

- It has criticised the MOD for insensitivity towards the problem.

- For downplaying the prevalence of the problem.

- For not being transparent.

- Being over secretive (made only two reports available for perusal by the committee out of seven).

- For not correctly perceiving the issue and devise appropriate strategy.

- For not taking corrective action in the crucial areas.

The committee has called for-

- Realistic assessment of the problem plaguing the Armed Forces and

- Reducing the stress levels causing psychological imbalance in the soldiers.

The committee emphasised on the need for-

- Statutory Provisions mandating the District Authorities to address the grievances and problems of serving Defense Personnel within a stipulated timeframe and formulating a mechanism for Central Monitoring System.

- Implementation of three recommendations to resolve the Familial issues; Providing family accommodation at the station of choice of personnel deployed in operational areas, imparting skill development and helping children in getting admissions to schools/Institutions of higher professional/ technical studies as also providing more Welfare Funds to Welfare Organisations of Defence Services so that they could extend necessary help to distressed families and counsel them.

- Making up the shortage of officers at the Middle and Junior levels in CI Environment and sensitising them.

- Rationalising the duration of employment of personnel in HAA and CI Environment.

- Removing the veil of secrecy and make the studies public in national interest and Improving the basic facilities(quality of life of the soldiers)

A similar view has also been expressed by senior Army Officers like Lt Gen Vijay Oberoi (Retd), Ex Vice Chief of the Army Staff and Brig Rahul K Bhonsle in their Articles "Fighting systemic changes are essential" [43]) and strike at the roots of Fragging. [44]

This study will enumerate on the above and also address some of the other issues, both at the macro/micro levels which need attention at the MOD and the Army Headquarters levels respectively. According to this Study, there are both systemic and non-systemic issues. The systemic issues need the attention of the Central Government/MOD while the Non-systemic issues could be dealt with by the Army.

Part 1- Systemic Issues

Reducing the Pressure on the Army. The Army is not only over stretched but it is also under pressure due to over load of work. There is therefore a need to reduce the pressure on the Army by taking the following measures:-

- **Need for National Aim and Strategy for LIC** . There is an urgent need for the formulation of a National Strategy of LIC like the USA has done for Iraq and Afghanistan. A workable Strategy, the comprehensive marshalling and utilization of

---

[43] Tribune, 31 Aug 2007.

[44] Indian Express, 04 Aug 2007

all available means/ resources to achieve national goals can flow from a National Aim. For example in J&K, such an aim implies a political solution to the problem. Unfortunately, a combination of lack of strategic thinking and poor coordination between various organs/ agencies of the Government precludes formulating an aim. But most importantly, domestic political compulsions inhibit the forging of a clear aim and strategy; yet without a clear national aim an inviable fundamental principle of any internal security campaign, it is not possible to formulate a strategy. The national aim and strategy would vary from region to region. For example, the aim and strategy for the NE Region (and for each state of NE Region) would differ from that of J&K. Based on the national aim & strategy, a political directive needs to issued to the Army Headquarter and all the Organs/Agencies of the State. The nation must no longer depend upon its strategy of "Attrition" but orient itself to pursue the strategy of economic development and security. Both activities must go on simultaneously. Based on the above, the Army Headquarters must formulate its own military aim and strategy and issue the same to the Army Commands as a directive. The directive would become the basis for conduct of all military operations.

- **Need for an Institutional Framework for Management of LIC:-** In the absence of a National Strategy and Directive, there is lack of co-ordination between various Organs/ Agencies. This precludes the employment of the full potential of the state to combat Insurgency or terrorism. The UPA Government had realised the need for an institutional look at internal security and had therefore appointed a separate Internal Security Advisor. Mr MK Narayan was originally assigned this task. However, consequent to his being appointed as the National Security Advisor, one is not sure what has happened to the group working on Internal Security.

LIC cannot be treated as another law and order problem. Militancy sponsored and abetted by external powers, is an act of aggression and requires different handling. The institutional framework and structure would require a separate organisation and dedicated staff at the National Security Council with representatives from all concerned Agencies. This may have to be replicated at the state levels. This staff must look not only at the Military aspects but also at the external linkages, internal, political problems, diplomatic, economic and intelligence aspects, work out solutions of these issues and coordinate their execution.

• **Resolve Dichotomy between Responsibility, Resources and Accountability**. Once the Institutional framework for the management of LIC is in place, then the adversial dichotomy between responsibility, resources and accountability presently misplaced will get resolved. This will help the Army to function without undue interference from the state leadership and define the relationship between the Army and CPO Forces with regards to their command and control.

• **Encashing the Operational Capability of the CPO Forces/ State Police.** A lot has been talked about but little done to upgrade the efficiency and operational capability of these forces. However, another effort has been made towards the modernisation and encashing their operational efficiency. The Government has allocated separate funds in the Budget 2008-09 for Internal Security

  • Rs. 11,486 crores. (Increase of 2.3% over 2007-2008)

  • CPO Forces Additional- Rs 4,041. 25 Crores over 2007-2008.

- Separate Rs 100 crores for developing critical infrastructure mainly Police Stations.

- CRPF, BSF, ITBP, CISF AND SSB- Rs 21, 715.25 crores. (+23%).

- Modernisation of state Police for purchase of Vehicles, Wireless sets and sophisticated equipment: Rs 1,340. 63 Crores.

- Training of Police in CI Operations. 210 Crores.

- It is hoped that this time full attention will be paid to equip and train the CPO Forces and the State Police to deal with

- Terrorism. This will provide great relief to the Army, which presently bears the main burden.

- **Raising of a "Third Force"- Internal Security Force**. The Rashtriya Rifles has done a commendable job – but it cannot remain a Para-Military Force in Army uniform for obvious reasons otherwise the Army will be viewed in different light. It is time therefore to re-organise the RR and raise a third force called the Internal Security Force (I &S Force). The nation will then have three types of Security Forces, namely, the Police at the state level, BSF and CPO Forces under the center (MHA) and the IS Force under the Army with two wings under separate DGs, separate uniform and insignia:-

  - IS Force (North East). Present AR (ASSAM RIFLES) to be redesigned.

  - IS Force (North and West). Re- organised RR.

  - The IS Force should train under the Army. The issue of budgeting would need to be resolved.

- This Force could also be tasked for Rear Area security during active hostilities.

- **Reduce Prolonged Employment of Army in LIC Operations.**

  - The fall out of the frequent call on the Army to assist the civil administration in restoring law and order situation are well known. The Army is well aware of its obligations in intervening in situations like that obtained in J&K and the NE. The compulsions of the Government are understandable. In that context, the Army has fulfilled its responsibility both in the NE and J&K.

  - But what has caused frustration in the minds of the officers and men of the Army is the inability of the Govt to politically resolve the issue. Having inducted the Army, the Govt seems to be happy with containing the problem. There appears to be no will on the part of the Government to go beyond 'containment. In fact, the very containment of the problem by the Army seems to have put the Government in a state of complacency. The Army therefore gets frustrated with the inaction of the Government (State and Centre) to initiate actions even when the level of violence has been brought down to an acceptable level of creating conditions of normally for the state Government to become effective resume control and carryout development work. What surprises the Army is that the state.

  - Government is more than willing to allow even expects the Army to carryout the development works. This is a startling example of flogging the willing horse little realising that the Army cannot set up Industries, provide employment and so on.

- It is high time that the Govt must revisit the concept of using force as a sub medium rather than an ultimate attempt to tackle insurgency/terrorism.

Restoring The 'IZZAT' of the uniformed man. The nation must take pride that our Armed Forces are a symbol of secularism, national integration, apolitical and unity. They have stood the test of time and maintained the freedom which was won by our freedom fighters. They therefore deserve to be honoured and given the **Izzat** they commanded at the time of independence but that has been eroded gradually. This is possible only if their "warrant of precedence "is restored and financial status enhanced by giving better pay and allowances.

Making up the Shortage of Officers. The Army has lived with the shortage of Officers at the junior level. Its adverse effects are apparent. Man management and welfare of troops is being neglected with serious consequences. It is therefore recommended that this shortage be made up expeditiously. The last route to making up this deficiency is to increase the intake of Short Service Commissioned Officers. But this avenue can be exploited only if certain conditions of service in this category of officers are amended so that they no longer feel cheated. They should be granted pension at the end of 14 years like it is being done for the soldiers retiring on completion of 17 years service. They should be extended the facilities of Ex Servicemen Contributory Health Scheme (ECHS). The best option is to grant them Permanent Commission on rendering 14 years of service.

- The Rashtriya Military Schools and the Sainik Schools are a rich nursery for the future cadre of Army Officers. The quality of candidates has the requisite knowledge and aptitude for the Army. Some of them have risen to higher ranks. The case in point is the ex chief of the Army Staff, General Deepak Kapoor, who is a product of Sainik School, Kunjpura.

Incidentally, his classmate, Chaudhary Bhupinder Singh Hooda is currently the Chief Minister of Haryana.

- The selection system also requires a review to ensure that some good candidates from the rural areas are not at a disadvantage because of not being fluent in English. The selection system must also ensure that candidates with inhuman qualities and those prone to anger with abusive behaviour are weeded out.

- The minimum service condition of 5 years be reviewed and one year service be reverted to apply for commission through Army Cadet College (ACC). 5 years is a long time to kill the ambition of prospective candidates and they reconcile to their fate of being PBOR.

Institutional Frame work for Redressel of the Grievances of the soldiers back home. The Zila Sainik Boards are inadequate to handle the grievances of the soldiers. There is a need to have people at the grass roots level that can follow up such cases with the local administration. One method could be to have a group of Ex Servicemen under a retired Officer(Col/Lt Col) in each district, pay them a small honorarium and organise the follow up of cases referred to the DC or SSP or any other district Official by the units. There will thus be an institutional framework to take care of this issue. There is also a need for the counseling of the separated families, especially those in the rural areas. (Enough has been said on the subject by the JPC and this issue needs no further emphasis).

Enhance Scales of Family Accommodation/ Separated Family Accommodation. (Enough emphasis has been given on the subject by the Standing Committee of Defence).

- The existing scales of married accommodation needs to be enhanced from 35% to 50%. All entitled officers and PBOR should be able to spend more time with their families when on peace tenure.

- Additional Separated Family Accommodation should be constructed in Astrik Military Stations of various regiments. This will facilitate better security for the separated families. They could avail the medical, canteen and other facilities like better schools for the education of their children. The scale of accommodation could be worked out by the Army based on the requirement.

- Families of Officers and PBOR could shift to this accommodation whenever the Unit moves out to field and revert to peace stations on return of the Units for peace tenures.

Improved scales of Pay and Allowances. The Sixth Pay Commission recommendations have not met the aspirations and expectations of the Armed Forces. This is evident from the fact that the three chiefs have taken up the matter with the Minister of Defence, Shri AK Antony who too concurs with them. Despite the Chairman of the Sixth Pay Commission trying to justify his recommendations, they were not satisfied and the case has been referred to the Ministry of Finance. While most of the anomalies have been resolved, the basic issue is that the pay scales are not comparable to the IAS, thus making no difference in the "Warrant of Precedence" and the 'IZZAT' of the soldier. If these anomalies are not addressed, then they may have serious consequences of low intake in the Armed Forces and more officers seeking premature release. It is recommended that in the better interest of the Armed Forces and this country, there should be a Separate Pay Commission for the Armed Forces and the veterans should be granted One Rank One Pension.

Part 2 - Non-Systemic Issues

The want to transfer the blame for all the ills to the bureaucrats and politicians would not be prudent. The Army has to apportion its share of blame. Therefore, there is an urgent need to do some serious introspection and soul searching. Above all, the Army should have the moral courage to accept the realities and take corrective actions to redeem the situation. In the context of the above, there are some core remedial actions required on certain issues within the realm of the Army (Non-systemic issues). These are contained in subsequent paragraphs. Some of the above issues have been projected but there has been a total lack of response. Other than these, there are other issues projected but are still pending with the Ministry of Defence.

Measures taken by Army. The Army too has taken several measurers to increase the awareness among the troops and minimize casualties due to suicides / fratricides. But these do not seem to be enough to arrest the menace of suicides and fratricides. A lot more needs to be done on the issues being mentioned here after.

Prevention of Stress, Suicides and Fratricides. The ideal would be to prevent Stress, Suicides and Fratricides. While stress cannot be eliminated, it can be reduced; Suicides and Fratricides can be prevented.

- **Stress.** Managing Stress is a command function. Commanders at all levels are responsible for it. For managing stress, the environment needs to be sensitised. In this context, the commander must look at himself and see that he is not the one who is a stress creator.

- **Suicides.** Though it is very difficult to identify a person who could commit suicide, but it is important that a person, who is under depression and has low tolerance is identified so that his problem is addressed and suitable measures taken to resolve the same. A person contemplating suicide displays peculiar behavioral pattern (Appears depressed, increased alcohol/drug abuse, non participation in group activities, talks about suicide/death and so on). Such a person should be counseled. Cases of attempted suicide have to be kept under vigilance.. This is where the problem lies. It takes away manpower resources to keep an attempted or likely case of suicide under guard. The best therefore would be to medically board him out. (*The best way to prevent suicides is to motivate a person to think that while he may think that he is one of the many in the world, but he is the only one for somebody in the world- for his family*)

- **Fratricides.** Similarly, a person with certain negative traits like being over sensitive, intolerant, impatient and aggressive, could be prone to losing his cool. Such a person could be easily provoked into killing his colleague or superior. Extra care is to be exercised while dealing with such individuals.

New Selection System for Officers and Recruits.

- The DIPR recommended in Jan 2007 to the Govt certain changes in the selection of Officers and Other personnel to tackle the problem of suicides and fratricides in the Army as told by Shri Manas K Mandal, Director, DIPR in New Delhi. The starting point is the right selection of officers and PBOR. The Psychological test for recruits was to be introduced from Oct 07 but still not implemented. (The process was to screen and detect cases of likely depression, suicidal tendencies and aggressive behavior).

- A new Selection system for officers is being formulated that will take around three to four years to develop and is expected to be implemented during the 11th year plan. The new selection system will stress on identifying positive traits and ensure detecting negative traits like depression and anxiety in the personality of the candidates. The emphasis will be on selecting candidates with holistic personalities who have good cognitive and intellectual capabilities. The new selection system is also geared to changes in warfare and focuses on equipping personnel to take decisions at the individual level. (Refer to an Article, Psychological Tests for selection of other Ranks- THE HINDU).

Leader –Led Relationship. The focus of refinement is the leadership and its relationship with the men. *It must be "Bottom to Top" approach instead of the traditional "Top to Bottom" approach*:

- **Higher Leadership**. If the senior officers want to bring about change, then they must become the change. The higher leadership is to be the embodiment of what leaders should be like-. They have to set an example of high integrity, truthfulness and uprightness both personal and professional. They must not only practice but also ensure that there is no concealing the truth, falsification, exaggeration, plagiarism, corruption and attempts at stealing credit. Unchecked corruption and unwholesome rivalry permeates every facet of our life and threatens the very foundation of service. They must practice what they profess and not exhibit double standards, particularly in seeking privileges. Since LIC is a war fought at the junior leadership, senior officers should desist over direction and over supervision thereby killing the initiative of subordinates. They must be more humane at resolving the problems faced by lower formations/units. Finally, they must devote more time to strategic issues like strategic intelligence, management of state leadership, media and the environment.

- **Middle Level Officers.** The middle level officers must learn to draw balance between their professional ambition and self induced pressure, leading to the flogging of troops. The CO has to be a buffer between the senior leadership and his command. He has to accomplish the task assigned to the unit, without passing down the pressure from top.

- **Strengthening the Junior Level Officers.** They must cultivate and practice the art of open leadership. The soldiers must be kept well informed-orders, briefings, debriefing, format talks and informal chats all help in awareness, understanding the overall situations, explaining the why of actions and the participation of the JCOs/NCOs. One must be accessible to the soldiers but without undermining the chain of command. One must empathise with theircommand, share in their privations, live hard and be seen to live hard. Laying continuous emphasis on operational preparations and training of command is also important. In an adverse situation, one must never allow the soldiers to take the law into their own hands, no matter how grave is the provocation. The challenge is to balance mission accomplishment; the safety and well being of thecommand and the preservation of good order and military discipline.

- **Junior Leaders.** The JCOs and NCOs can no longer afford to be parochial and unfair, especially in the distribution of duties or granting leave. The JCOs and NCOs should be shown due respect, taken in the loop of decision making and held accountable for lapses in their respective commands. They should not be ignored, bypassed and made to feel irrelevant.

Sensitising the Leadership

- The officers may be as caring and sensitive to the sensitivities of the soldiers as their predecessors but do not seem to keep in mind that they are dealing with a new man in uniform. They must understand that the concept of welfare has changed. Leave, authorised rations and promotion cadres have been augmented like due care, respect and need for polite but firm handling. Similarly, the society has given a more impatient, intolerant and aggressive man to the Army, Therefore, the soldier of to day is more sensitive to being humiliated abused or facing injustice- not that these were not decried before.

- Not withstanding the pressure of work, the Officers cannot absolve themselves of ignoring the welfare of their men. They have to pay more attention to the quality of food, living and clothing of men, thereby creating a conducive environment at the COBs (Company Operating Bases).

- Provision of colour TVs and DVDs cannot be a substitute for close interaction—the officers, JCOs and NCOs must interact with the men without intruding into their privacy. In exercising interpersonal relations, the officers, JCOs and NCOs need to exercise restraint when angry and not hurt the sensitivities of the soldiers. The soldier must not be humiliated/ abused/ harassed or admonished in public.

- There is a need to start including teaching of the subject of Man-Management in the present context. The exposure must commence at IMA and thereafter, at every leadership course JC, SC and HC. Similarly for the JCOs/NCOs, it must be in the curriculum of the junior Leader course.

- A special column could be included in the ACR Form, to reflect on this aspect.

Women Officers. The Women officers are an asset to overcome the shortage of officers. The mindset of a gender bias towards them must therefore change so that they can settle down well in the new environment. They must also be employed, keeping in view their qualifications. Above all, they must not be unduly harassed. On the other hand the women officers must be prepared for the hardships of the Army life and not expect ideal administrative arrangements.

Working Environment. The men (All Ranks-Average-80%) are under tremendous pressure due to heavy work schedule. It is because in LIC, Operations are generally carried out on mere speculation rather than based on intelligence resulting in the flogging of troops. In peace stations there are far too many training, sports and social events that keep the troops quite busy and they do not get time to be with their families. Besides, whether in peace or field, too much time is spent on rehearsals at various levels to achieve perfection. In order to give adequate time for soldiers for rest/ personal administration the following needs to be done; more importantly give them quality time to be spent with their families.

- In Peace, there is a need to outsource jobs which can be executed by other agencies and also reduce the number of training, sports and social events. In CI Operations, the men chase the militants and in peace they chase the events.

- In LIC, operations should generally be carried out based on some intelligence.

- Review the existing policy of units on peace turnover carrying out field firing and collective training at various levels, giving the men no respite.

- Minimum time should be spent on rehearsals.

- Machines should be used where possible.

There are certain commanders who generate unnecessary work. Most of them are ambitious and try to flog the men to show better results than their counter parts.

Quality of Life. There is a need to understand the correct concept of welfare of men. Though the Indian soldier can brave a lot of hardship; live under improvised accommodation, would not complain about the quality of food and will not demand too much of recreational facilities but the Army must not ignore these factors just because the men are uncomplaining. Every effort must be made to make their quality of life better. The staff at higher head quarters must spend more time improving the logistics and transit facilities instead of being tied down to Power point presentations.

Empowering the soldier. The soldier feels let down when fighting the militants who have better communications and weapons. The soldier does not find a level playing field. This is very demoralising and leads to stress. There is an urgent need to provide the latest weaponry and specialised equipment to the soldiers so that they can confidently confront the militants. If the soldier is expected to perform, then his performance capability needs to be enhanced accordingly.

Psychiatrists and counselors. In order to ensure forward treatment, there is a need to enhance the number of Psychiatrists and clinical Psychologists, both in field and peace. A minimum of one Psychiatric centre with adequate supporting staff (one Psychiatrist, two Psychiatric Nurses and four Psychiatrist Nursing Assistants) should be authorised at each Divisional/ Area Headquarters for forward treatment. This Centre should maintain the necessary data bank for periodic analyses. Similarly, besides imparting ad hoc training to the unit personnel, there is a need to have qualified Counselors at the Divisional level. These Counselors should visit units and talk to troops on Stress related problems. (Refer to an Article, Suicides by Soldiers: Army to send more shrinks to J&K-The Tribune, 02 Nov 06).

Miscellaneous Issues.

- **Mis-use of AFMS-10 .**The AFMS-10 is being misused both by the officers as well as the men to get rid of difficult cases or get away from the harsh environment. Such cases of misuse must be identified and punished and if necessary the AFMS-10 form should be reviewed.

- **Restricted use of TV and Mobile.** The mobile being allowed as an interface with the families is a double edged weapon. It is probably creating more problem for the men than resolving them. Similarly, it has been observed that troops spend more time watching TV than taking rest or attending to their personal administration. While it would be preposterous to totally ban the use of the TV or the mobile; it would be equally incorrect not to regulate their use.

- **Alcoholic and Debt cases.** These being the major causes of mental disorders and subsequent suicides, need to be kept under strict watch in the units.

- **Army Ethos.** There is a need to change our rigid attitudes. The young officers need to be listened to. They are quite enlightened and professionally well informed. Similarly, the jawans must be given opportunities to express their views and their emotions should not be suppressed.

- **Tenures in LIC.** The troops are generally happy with two years tenure in LIC area. But the process of Induction and de-induction with consequent orientation and de-orientation after every five or six years, causes stress. This needs examination at the Army Headquarters. Secondly, the problem relates to those who are posted to RR. Some of them find themselves posted to RR soon after the unit reverts to a peace station after its tenure of CI Environment. The problem also relates to those who are not relieved on completion of their tenure with RR.

In LIC Environment only.

- **Pre-Induction Training.** In LIC Environment, a soldier fights a war with constraints; deal with a hostile population, an adverse media and unsympathetic Human Rights Organizations. The Corps Battle schools (CBS) are well equipped and their training syllabus is correctly structured to prepare troops for the tasks that would be required to carry out on induction. In fact, 15 Corps has gone to the extent of re-creating a mock village to give officers and men a very good and realistic feel of the areas where they would operate. The duration of training varies from four to six weeks which is extendable if so desired by the commanding officers and/ or formation commanders, who know the strengths and weakness of their outfits the best. Therefore, while the tactical orientation is good; regrettably, the Pre-Induction training is not realistic and does not prepare the soldier psychologically to fight in an environment peculiar to LIC. The Pre-induction training must therefore be realistic to orientate the soldiers' mind to this new environment and the rules of conflict- all locals are not militants and majority are innocent; the adversary with the gun is not an enemy, the people may be hostile, the media may be scathing in their attack and the Human Rights Organisation may seem to be sympathetic to the militants.

Freedom of action in executing operational tasks. There is a tendency for higher commanders to conduct operations by applying the template of either similar situations or their earlier experience obtained else where. Such commanders must remember that CI operations are generally conducted at unit / sub unit level. It is therefore a junior leader's battle and he must be allowed full freedom in execution of the tasks assigned to him. Commanders at higher echelons should not chase him. Instead, depending upon the level of command, the senior commander must concentrate on either shaping or managing environment.

The Pressure of Numbers Game. The phenomenon that has profound psychological impact on the troops in CI Operations is the body count or "The Kills and Recoveries"- numbers game and the negative competitive spirit it generates amongst the units for one up-man ship. It also leads to unhealthy practices like fake encounters. Where does it start? Without debating the issue; whether the pressure is generated from the top or by ambitious commanding officers, it needs to be emphasised that no one irrespective of the rank, should be allowed to flog the troops to just achieve higher results. It is inherent in the task and needs no further goading for it.

Zero Error Syndrome. In CI Operations, senior officers need to exercise "Patience and Tolerance". Events could get out of hand. If all commanders want results, then they must also be tolerant enough to accept reverses, if any and failures too. Though accolades are showered on the same company commander and young officer when they get results- incidentally they are the one who get results- whatever be the cause, mistakes are seldom tolerated, then why is it that they are taken to task when in their exuberance, they make mistakes? Such mistakes even when made inadvertently, are not condoned- Not many care to hold their hand. Mistakes like co- lateral damage or killing of an innocent are the common fall outs of CI Operations. In such cases, the organisation must fight for him. This is where the basic meaning of responsibility needs to be recalled- 'When everything goes bad, I did it-If anything goes some where good, then we did it-When everything goes good , you did it.

Fear and Uncertainty. In CI Operations, there is always uncertainty about an unexpected attack and the fear of being killed. The biggest concern of a soldier is the concern about his family, in case of his death. If such a tragedy does take place, then, it is the duty of the state to take care of the family and the children.

Winning the Hearts and Minds. The civil population is the centre of gravity in CI Operations. Therefore, winning the hearts and minds of the local population is the key to the success of CI/CT operations. It is with this aim and with a view to shaping or influencing the environment that various commanders devise ways and means to get close to the population. While the shaping of the environment should be attempted at levels of Corps and above, the influencing should be practiced at the divisional and lower levels. It is in this context, that Sadhbhavna started as an effort in this direction. There was no template or a predetermined design of what measures were to be undertaken to win the hearts and minds. That was to be based on and taking into account local conditions. However, over a period of time Sadhbhavna became an event of one up man ship and units started doing what pleased their superiors more than what the locals desired. Thus a fine initiative was ultimately viewed by the troops as Sad-Bhavna. Sadhbhavna, is now resented both by the officers as also the men and equally by the locals. It is important to win the trust of the locals while carrying out civic programme. Not withstanding the fact that the works are executed through the contractors, there is still some involvement of the unit personnel. It is thus important that while carrying out such civic programmers, resources of fighting units are not put to use. Special task oriented teams designated for this purpose preferably from outside formations should be employed.

Management of Stress. To manage Stress, remember-

"These days, it seems that all of us are too serious. Understand the link between your expectations and frustration level. Whenever you expect something to be your way and it isn't, you are upset and you suffer. But when you let go your expectations, when you accept life as it is, you are free. To hold on is to be serious and uptight. To let go is to lighten up."

- Richard Carlson

Conclusion

Combat stress and its attendant problems of suicides, fratricides and other psycho/neurotic disorders are inherent in an LIC environment; there is therefore no cause for worry but it is a matter of serious concern. While the problem cannot be totally eliminated but certainly it can be minimised. Every human life saved is worth the effort.

The issue is a harsh reality and no lip service will do if this has to be tackled. It is to be recognised as a national issue and the Army cannot be left alone to struggle on its own.

- Firstly, the political leadership must understand the vital importance of good governance in ensuring that people don't reach the stage of taking to violence to seek solution to their grievances.

- Secondly, when faced with the problem, there is a necessity for a National Strategy and an institutional framework to deal with it.

- Thirdly, the nation will need to revisit the very concept of force as a sub median rather than an ultimate solution to the problem.

- Fourthly, in the present context, while the government handles the systemic issues like shortage of officers, pay and allowances and mechanisms for the redress of the grievances of the soldiers by the state governments, the Army will have to take a hard look within to improve the intake of officers and men and their training, improve empathy between the officers and the men - officers to be more sensitive and caring and ensure implementation of basic drills and procedures in the unit through strict discipline.

- Lastly, the studies carried out earlier and this Study will serve their purpose only if their recommendations are given a serious look and implemented with sincerity. The effort of the Study will otherwise be futile if it is also consigned to the archives of the library.

There is enough knowledge available with the government and the Army with regards to the causative factors of stress. What is lacking is action that must be put into motion, with vigour and sincerity.

In the end, we must remember what **Richard Gabriel** has noted:-

"Nations customarily measure the cost of war in dollars, lost production or the number of soldiers killed or wounded." But rarely do military establishments attempt to measure the costs of war in terms of individual suffering. "Psychiatric breakdowns remain one of the most costly items of war when expressed in human terms" Indeed, for combatants in every war fought in this century, there has been a grater probability of becoming a psychiatric casualty than of being killed by enemy fire.

CHAPTER 14 - THE WAY AHEAD

Introduction

Having arrived at some conclusions, the larger issue is: What **Next**? Well, the answer to this question is simple: begin with the beginning- that is start with the process from the time the orders for induction into an LIC- environment are received and include all activities to be undertaken prior to the induction, post the induction and the De-induction.

While laying down the actions to be taken, it must be remembered that LIC environment is stressful: implying it generates stress. Stress as has been explained earlier is the mother of all other maladies. It makes a difference between a successful and an unsuccessful tenure of a unit. Therefore, the prime objective of any preparations for tenure in an LIC environment is to train the unit personnel in containing stress within acceptable limits; restore it when it becomes temporarily disruptive and increase tolerance levels to stress.

The actions to be taken are discussed as under-

- Section 1- Pre- Induction Stage.

- Section 2-Induction and Combat Stage.

- Section 3- De- Induction Stage.

Section 1- Pre-Induction Stage

Preparatory Actions

This phase is vital and includes preparations peculiar to LIC environment prior to the unit moving to the Corps Battle School.

- **Psychological Screening.** The first step is to arrange for the Psychological screening of all its personnel. The aim of such a screening would be to identify cases having suicidal tendencies or those who are prone to alcoholism. But most importantly to identify those who are mentally weak and likely to succumb to pressure under adverse situations/environment.

- **Redressal of Family Problems/ Separated Families.** The unit must ensure proper settlement of separated families and send others to their homes. Problems of personal/ domestic nature must also be settled with the help of the State/ District administration. All personnel must be encouraged to update their records with regards to the Next- of- Kin (NOK), wILL, Insurance Policies, latest home address/ telephone numbers for contact and finalise Power of Attorney where applicable.

- **Deficiency of Weapons/ Equipment and Clothing.** The unit must make up all deficiencies especially in weapons/ equipment including spares and batteries. All unserviceable clothing items must be replaced including the boots. Extra pair of combat dress must be arranged as they have a heavy wear and tear.

- **Cohesiveness of subunits.** The deficiency in manpower of various sub units must be made up and these must be organised into cohesive groups.

Pre- Induction Training

This involves the orientation of the unit at the Corps Battle School (CBS) for the impending task. The Army already has the system in place to impart the necessary training at the CBS. However,from the interaction with the retired officers, JCOs and ORs , it is learnt that while there is enough stress on the tactical aspects but there isn't enough time to cover some of the other subjects which are job related. These are given in succeeding paragraphs.

- **Mental Robustnes.** In CI operations, there will be fast paced events; frequent deployment and redeployment, operating under difficult terrain and harsh climatic conditions, deprivation of rest and sleep ,and irregular meal timings. Above all, there will be most adverse situations. It is therefore imperative that officers and men are mentally prepared for such situations. Putting them through adverse situations would build up their mental robustness.

- **Yoga and Meditation.** Individuals should be put through a capsule of Yoga and Meditation. Since there is no time for regular physical training, this would strengthen their faith in the Almighty and improves their physical well being. It will also inculcate stress- coping skills.

- **Firing Standards.** Besides improving the general firing standards, men must be practiced in firing from the hip. There should be due emphasis on Fire discipline.

- **Sleep Discipline**. It must be remembered that there will be fatigue and loss of sleep during combat. During continuous operations, fatigue caused by sleep is a major source of stress. Therefore, enforcement of rest and work schedules must begin early during the Pre- Induction training. There will be unscheduled and infrequent breaks in combat. Sleep must be allocated or supplied like rations and water, equipment and ammunition. Relaxation exercises and quick naps complement sleep. These exercises must be used to supplement sleep and relaxation or serve as an aid to help soldiers rest under difficult circumstances. It must be remembered that only 5 hours of sleep over a period of several days will accumulate a significant sleep debt.

• **Management of Media.** Media is very active in CI operations. It is therefore imperative that officers, at least at the company commander level are trained in briefing the media. Our officers are media shy. Reluctance to brief the media on time will go a long way in giving the correct story and counter act the false propaganda by the militants/ Terrorists.

• Troops must be briefed about the task ahead (aim of operations) and why they are being inducted, dealing with the local population, response to casualties, importance of causing minimum co- lateral damage (fighting with constraints) and so on.

Section 2- Deployment and Combat Stage

This phase is generally associated with the normal stress of having been inducted into an LIC environment, generates thoughts about survival and performance under fire. It will also create the fear of the unknown, unexpected attack and soldiers would start worrying about the future of their families in case of their being a casualty. All these factors coupled with ad hoc living conditions would add to their anxiety and fear leading to normal stress. This is the time for the officers and JCO's to assuage their feelings and put them at ease through constant interaction with them.

A period of approximately 15 days overlaps between the out going and the incoming units. During this period, there is familiarization of the area of operations and tasking. This period is vital to the further success of the unit and must be utilised fruitfully. The other aspects requiring attention are as follows-

• **Flow of Information.** Operations should usually be based on information. It is therefore important to establish an Intelligence Grid for the flow of information. Besides, there is a need to keep the troops informed. Lack of any information leads to rumours and a cause for anxiety. The men must be asked to write to their homes or speak to them and intimate about one's well being.

- **Integration of New Members.** The arrival of young soldiers requires a detailed integration plan. They must be associated with somebody who hails from his district/ village. They must be given an orientation to the task and stress coping skills. They should gradually be blooded into action and imparted on the job training.

- **Religious Teacher Support.** The RT is an integral part of the unit. Full use must be made of skills and professional training to focus on prevention of mild and moderate combat stress reactions. The RT should also provide comfort, assurance and encouragement to the soldiers through religious discourses so as to restore faith in God and religion.

- **Rest and Recreation**. Sub units in rotation must be pulled out periodically for rest and recreation. This period should be a period of unwinding through listening music, watching TV/movies and writing letters home. Adequate time for personal hygiene and facilities are psychologically valuable in combat. The officers must attend to any personal or domestic problems of individuals.

- **Kill/ Capture Militants**. The men must understand their mandate; that is to restore normalcy and ensure a peaceful environment. In so doing, they would be expected to remove the fear of the gun. In this process, it would be inherent in their task to kill or capture militants. But it needs to be understood that while elimination of the militants is important, obtaining the number of kills is unimportant.

- **Attitude towards the Local Population**. The local population is the centre of gravity. They would appear unfriendly and non co-operative. But that is because they are under the shadow of the fear of gun of the militants. Therefore, every effort must be made that neither should any innocent be killed in an encounter nor undue damage caused to the property But the most vital aspect to be observed is to maintaining the privacy and dignity of the locals. Equally, they should neither be disturbed nor harassed at odd hours unless inevitable.

- **Response to Casualties.** While conducting CI operations, casualties are inevitable. After all when militants are killed, the militants are bound to inflict casualties on own troops. While it would be natural to be angry on the loss of a comrade, there is no need to go berserk, as the possibility of getting killed or maimed cannot be ruled out. The men must be prepared for it. The important issue is that if it happens, then the man must have an assurance that the Nation/the State would take care of the deceased family. Any unit commander who is not prepared to accept casualties is likely to set in a defensive -mentality in his men and they would avoid an encounter with the militants. If that happens, then the unit is likely to end up with higher causality rates.

Stage 3- De- Induction Stage

This stage is as important as the other stages. The men would require time to de-link from the LIC environment. They would continue getting dreams and night mares of the incidents. They are likely to feel boredom and lack of sleep.

While there has been no study on Post Traumatic Stress Disorder (PTSD) in the Indian Army, cases of this malady cannot be ruled out as is evinced by a Retrospective study of all sailors returning from deployment in Iraq and Afghanistan, between 2006 and 2008. The study showed increased probability of PTSD by 6.3 and 1.6 percentage points respectively compared to those deployed on ships. There is a need to carry out a study on this aspect.

It must be remembered that troops reverting to a Peace station eagerly look forward to a break from the monotonous routine of patrolling, ambushes and raids. They aspire to be united with their spouses and children; spend some quality time with them. But what generally happens is that the atmosphere of the field is carried forward to a peace station. The troops soon start preparing for the visit of a series of senior officers and the usual Geru- Chuna starts. Soon before the unit settles down, a Training Directive is handed over containing the Training/ Sports calendar. It is not long before the unit is caught up in the vortex of Field Firing, Battle inoculation (which they don't really require) and Collective Training schedule besides preparing for the Sports competitions. The tragedy is that the men get inadequate time for rest and sleep; and more importantly to be with their families, resulting in continuity of Stress and Suicides. In order to give a feel of a Peace Station, the following is recommended:-

- There should be no visit of any senior officer for at least one month so that the unit gets time to settle down, men go on leave and bring their families.

- The unit must be spared from participating in any collective Training or Sports Competitions for a minimum six months.

- During the subsequent year, the unit should not get involved in more than one Divisional level/ Corps level event.

- Adequate married accommodation should be made available so that each man is able to spend at least two years of peace tenure with his family.

- The concept of Family Welfare Centre (FWC) and AWWA need to undergo a change; should be oriented towards events that are beneficial to the families.

Conclusion

The cycle of a unit earmarked for tenure in LIC environment, commences with its preparations and ends with the process of the men unwinding themselves and settling down for Peace tenure; to be ready for tenure in a similar environment. This process of Induction- De-Induction and Induction again is most stressful unless carried out with care and proper planning for each stage.

Bibliography

Books

- Stress and Coping by Dr D M Pestoneji.
- Kashmir Diary by Maj Gen Arjun Ray.
- Tackling Insurgency and Terrorism by Maj Gen Samay Ram (Retd) UYSM, AVSM, VSM.
- Stress Management coping with Military Situation by Kiran S Bhan.
- The Wounds of War- The Psychological Aftermath of combat in Vietnam by Handin H,Hass,Ap.
- Handbooks on Understanding Stress, Stress Management and Occupational Stress (I-III) issued to the Army.

Studies and Reports

- Evolving Medical Strategies in LIC by Lt Col Ajay Dheer et al
- Medical Journal Armed Forces India, 2003.
- Psychological effects of LIC Operations by Lt S Chaudhary et al.
- Indian Journal of Psychiatry-2006: 48:223-231.
- DRDO Report-DRDOconducts studies on-Stress. www.indiadefence.com/reports-3626
- DIPR Report-Occupational Factors, No 1 Cause for Suicides in the Army by Ravi Sharma THE HINDU, 08 Aug 07.

- Neurosis among Armed Forces Personnel by Maj C Dhir , Col A Bannerji (Retd), Col S Chaudhary(Retd) and Brig Z Singh. Medical Journal Armed Forces India,2008.

- Changing Pattern of Alcohol Abuse in the Army before and after issue of AO 3& 11/2001. Col S Saldhana, Col S Chaudhary, Surg Capt A A Pawar, Surg Capt VSSR Ryali, K Srinivasan, M Sood and Maj HK Bedi-.Medical Journal Armed Forces, India,2007.

- Extracts of study by 92 Base Hospital on Inter Personal Relations.

- Dissertation at NDC on Stress Management in the Army by Brig (now) Maj Gen SK Bhardwaj.

- Psychological Aspects of CI Operations by Col DS Goel, Combat Journal. Apr, 1998.

- CLAWS Research Paper on Challenges of Management and Combat Stress in LIC Environment (Forthcoming).

- Seminar on the Sixth CPC, held at the USI in 2007.

- Extracts of Seminar on J&K- Contours of Future Strategy held in CLAWS on 02 Jun 07.

- Analyzing the US Military casualties in Iraq and Afghanistan- Grand Strategy. www.grandstrategy.com

- 31st Report of the Standing Committee on Defence on Stress Management in the Armed Forces, Oct 2008.

Magazines

- USI Journal- 1992, 2004, 2007, 2010.

- Combat-1992,2000·

- Medical Journal Armed Forces,India-1996&2002

- Defence and Strategic Affairs-19 Jun 2007.

- Indian Defence Review-2003.135

Articles

- Fratricide Incidents in the Army by RSN Singh, Indian Defence Review2003

- Stress Management in CI Operations by Maj R Rajan-THE INFANTRY(INDIA),2004

- Adverse Psychological Effects of CI Operations by Maj P Badrinath-USI Journal -2003

- Saturday Extra- The Tribune.

- Press Information Bureau, Government of India Press Release-07 Mar 09

- Psychological and Motivation Factors affecting the soldiers by Lt Gen S C Sardeshpande, Combat Journal, Apr 1988.

- Officer- Man Relationship by Lt Gen Satish Nambiar (Retd) ex Director USI-South East Asia Defence and Strategic Review- Sep 2007.

- The Rashtriya Rifles by B Bhattacharya- Bharat Rakshak Moniter Vol 3(2) - Oct 2000.

- Women in the Army by Vipin Agnihotri..

- Military Pay Packet by Lt Gen Vijay Oberoi- Hindustan Times, Chandigarh, 25 Apr 07.

- Extacts of Fratricides and Fragging from Wikipedia, the Free Encyclopedia.

- Extracts on Suicides from Wikipedia, the Free Encyclopedia.

- Fragging- Panel flays Ministry for inaction by Rajat Pandit- Sunday, Times of India, 19 Aug 2007.

- Fighting Fragging –Systemic Changes are Essential by Lt Gen Vijay Oberoi- the Tribune, 31 Aug 2007.

- Strike at the Roots of Fragging by Brig Rahul K Bhonsle- Indian Express, 04 Aug 2007.

- Officers leaving Forces despite Pay Hike Promise-rajatpundit@times.com, 30 Aug 07.

- Psychological Tests for selection of Other Ranks of the Army-The HINDU.

- Sensitive Leadership Key to Arresting Army Suicides by Girja Shankar Kaura- The Tribune,09 Jul 07.

- Fragging: Humiliation, biggest trigger-Times of India, 09 Jun 07.

- Prolonged anti insurgency operations taking toll on jawans – Times of India, 11 Jun 2007.

- An Officer and a woman- Times of India,03 Jul 07.

- Fighting Stress- Sat Extra. The Tribune, 06 Jan 2007.·
Suicides by soldiers- Army to send more shrinks-The Tribune, 02 Nov 06.

- Reducing the Work Stress Among Army Personnel-Press Information Bureau, Government of India,07 Mar 2007.

- The Indian Army: Crises within- India Together, 13 Sep 2007.

5. Interaction with Retired Service Officers through questionnaire.

6. Internet.

www.ingramcontent.com/pod-product-compliance
Lightning Source LLC
Chambersburg PA
CBHW070809300326
41914CB00078B/1915/J